Computational Biology
AND
Genome Informatics

Computational Biology
AND
Genome Informatics

Editors

Jason T L Wang
New Jersey Institute of Technology, USA

Cathy H Wu
Georgetown University, USA

Paul P Wang
Duke University, USA

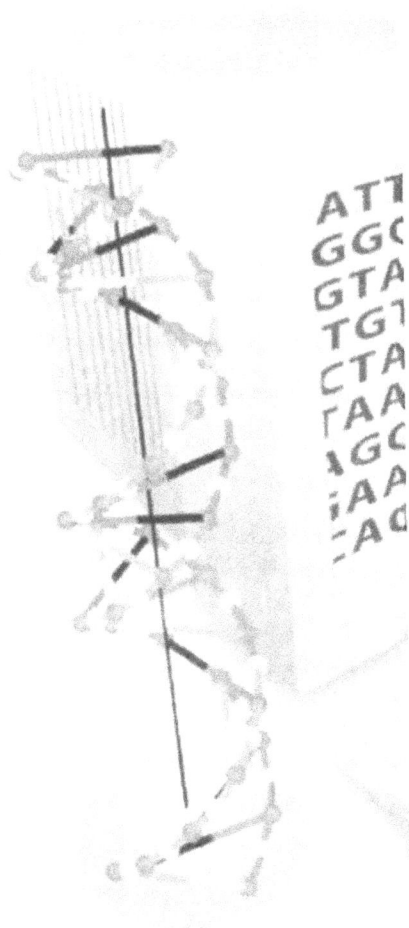

ATT
GGC
GTA
TGT
CTA
TAA
\GC
iAP
=AC

World Scientific
New Jersey • London • Singapore • Hong Kong

Published by

World Scientific Publishing Co. Pte. Ltd.

5 Toh Tuck Link, Singapore 596224

USA office: 27 Warren Street, Suite 401-402, Hackensack, NJ 07601

UK office: 57 Shelton Street, Covent Garden, London WC2H 9HE

British Library Cataloguing-in-Publication Data
A catalogue record for this book is available from the British Library.

COMPUTATIONAL BIOLOGY AND GENOME INFORMATICS

ISBN-13 978-981-238-257-3
ISBN-10 981-238-257-7

Preface

In the post-genomic era, we now have an unprecedented view of the genome of many species as well as new views of how biological processes occur. The availability of genomic and genome-scale information is changing the way biologists work and revolutionizing the way biology and medicine will be explored in the future.

To fully realize the value of the data and gain a full understanding of the genome and the proteome, advanced computational tools and techniques are needed to identify the biologically relevant features in the sequences and to provide an insight into their structure and function. Systematic development and application of computing systems are also needed for analyzing data to make novel observations about biological processes and to model biological systems with high accuracy. A large amount of data must be stored, analyzed, and made widely available to the scientific community.

This book contains articles written by experts on a wide range of topics that are associated with the analysis and management of biological information at the molecular level. It contains chapters on RNA and protein structure analysis, DNA computing, sequence mapping, genome comparison, gene expression data mining, metabolic network modeling, and phyloinformatics. It is addressed to academic and industrial researchers, graduate students, and practitioners interested in the computational aspects of molecular biology. The highly interdisciplinary nature of research in this area is providing a fruitful ground where a variety of ideas and methods come together. This volume is a sample of some of the major techniques currently in use in this cross-cutting field.

The book is the result of a two-year effort. We thank the contributing authors for meeting the stringent deadlines and for helping to create the index entries at the end of the book. Special thanks go to Hansong Sara Liu for assisting us with Microsoft Word software and other issues in the preparation of the camera-ready copy of this book. Finally, we would like to thank Yubing Zhai and Ian Seldrup of World Scientific Publishers for their assistance.

Contents

Chapter 1

Exploring RNA Intermediate Conformations with the Massively Parallel Genetic Algorithm

Bruce A. Shapiro, David Bengali, Wojciech Kasprzak and Jin Chu Wu

1.1 Introduction

The bioinformatics revolution has led to an exponential increase in the availability of data on gene location, expression, and function for thousands of species. In the midst of this eruption of data, however, time and resources are often lacking for the analysis of information beyond that encoded by sequence alone. While proteins are the traditional candidates for detailed structural analysis, RNA secondary and tertiary structural studies remain crucial to the understanding of complex biological systems. The RNA structure-function relationship list is quite long. Structure and structural transitions are important in post-transcriptional regulation of gene expression, intermolecular interaction and dimerization, splice site recognition, and ribosomal frame-shifting to name a few contexts. The ribozymes constitute a class of RNA molecules whose sequence exists primarily to define their structural and enzyme-like properties. The RNA folding problem clearly is a significant venue for the use of computational approaches. As with most such

1

applications of high-powered computing, the problem of RNA structure determination is a difficult one. The number of secondary structures possible given a particular sequence varies on the order of 1.8^n for a sequence of n nucleotides. Traditional approaches to the problem are numerous and varied. A wide range of biochemical and biophysical assays may be used to examine RNA secondary and tertiary structure. These assays generally search experimentally for the consequences of sequence and structure within a molecule, probing for accessibility to enzymes, calculating optical absorbency, or measuring electrophoretic migration rates over a temperature gradient. A given structure generally is verified through phylogenetic analyses, searching among members of a family for compensatory base changes that would maintain base-pairedness in equivalent regions. All of this fairly direct data often is supported, or at times even replaced, by theoretical structure calculations. The most familiar variety of these are derived from dynamic programming algorithms (DPA) such as MFOLD [Zuker, 1989], and which search for a molecule's thermodynamically optimal structure, as well as a series of suboptimal structures. When the object is secondary structure, that is, a structure that can be defined as a list of base-paired and single-stranded regions (stems and loops), thermodynamic calculations are straightforward. Stems tend to stabilize a structure and most loops tend to destabilize it, and the energies of these stems and loops are additive. Thus, a search for biologically relevant structures can be driven by the assumption that a molecule will tend to fold spontaneously into structures that minimize its global Gibbs free energy with respect to the unstructured state. A recent version of the dynamic programming approach to energy minimization has been able to include H-type pseudoknots and some basic tertiary structure energy contributions at the cost of moving the algorithm to $O(n^6)$ time [Rivas and Eddy, 1999]. By removing pseudoknot considerations and shifting the more precise tertiary structure energy calculations for multibranch loops to a post-processing reordering phase, this algorithm runs in $O(n^3)$ time [Mathews *et al.*, 1999]. Searching experimentally and theoretically for these equilibrium structures, either optimal or suboptimal, however, is often insufficient. The biologically functional state of a given molecule may not be the optimal state, and how, then, does one determine the relevant suboptimal structure? A structured RNA molecule, moreover, is not a static object. A molecule may pass through a variety of active and inactive states over its lifetime, due to the kinetics of folding, to the simultaneity of folding with transcription, or to interactions with extra-molecular factors. A molecule may become trapped in a local energy minimum with a high activation energy barrier to surmount before reaching a more stable state. How can one begin to approach the analysis of such a moving target, a target with a vast and highly combinatorial n-dimensional structure/energy landscape over which it may travel? Methods developed using a massively parallel Genetic Algorithm (GA) optimization approach have proven highly amenable to exploration of such RNA secondary structure folding pathways. This algorithm was designed using the same basic considerations as the dynamic programming algorithm; that is, with thermodynamic calculations to optimize the global free energy of an RNA molecule. As such, it is reasonably successful at efficiently finding optimal or near-optimal equilibrium structures, including pseudoknots, given a particular sequence. The properties of this massively parallel, iterated, stochastic algorithm, however, have revealed themselves to be ideally suited to the problem of predicting the dynamic folding process of a given molecule as well. In addition, the algorithm allows for the incorporation of some types of experimental data, allowing it both to verify and to predict the outcome of experiments under known conditions. The Genetic Algorithm operates on a population of thousands of possible solution structures, evolving them toward thermodynamic fitness. It may be run multiple times and in multiple phases. STRUCTURELAB, an interactive RNA structure analysis workbench, has proven indispensable in analyzing the large quantities of data generated by such use of the GA. In particular, use of Stem Trace, a STRUCTURELAB component for abstract graphical comparison of RNA secondary structures, has given great insights into a variety of RNA structural issues, including that of folding pathway exploration.

1.2 Algorithmic Implementation

The massively parallel Genetic Algorithm is a member of a class of algorithms that use the principles of evolution to optimize a parameter within a population of possible solutions. In this case, the parameter is free energy, but the optimal structure is not the only consideration. The intermediate results within the population and the pathway followed by the algorithm to reach its final solution are equally important. Still, the basic operators are as one would expect: mutation, selection, and recombination. The basic procedure is as follows (details on each step follow in the text):

```
 1 Generate stem pool
 2
 3 Fill structures in each Population Element randomly
 4
 5 while (not converged) {
 6         for each (Population Element), execute in parallel {
 7                 Find neighbors
 8                         Ranked-select and store two parents from self
                         and neighbors
 9                 Create two empty children
10                 Mutate children based on current structure size
11
12
13                 Crossover {
14                         for each (stem in parent 1) {
15                                 Distribute into one of the children
16                         }
17                         for each (stem in parent 2) {
18                                 Distribute into one of the children
19                                 if (stem conflicts)
20                                         if (probability == true)
21                                                 Try to peel stem to fit
                 into structure
22                                         else
23                                                 Discard stem
24                         }
25                 }
26
27                 Replace self with better child
28         }
29
30      Output intermediate data
31      Calculate z-score and convergence
32      Increment generation
33 }
```

Population

The population itself is made up of an array of elements, where each population element (PE) represents one structure. A structure in this case is uniquely defined by a list of base-paired regions, or stems. Thus the "chromosome" for a PE is this list of stems, which may be altered by the basic GA operators. The population is arranged in a two-dimensional grid, where each PE can communicate directly with its eight nearest neighbors (see Figure 1). The grid is wrapped toroidally, so that the northern neighbor of a PE on the northern "edge" of the array is located on the southern "edge" (the

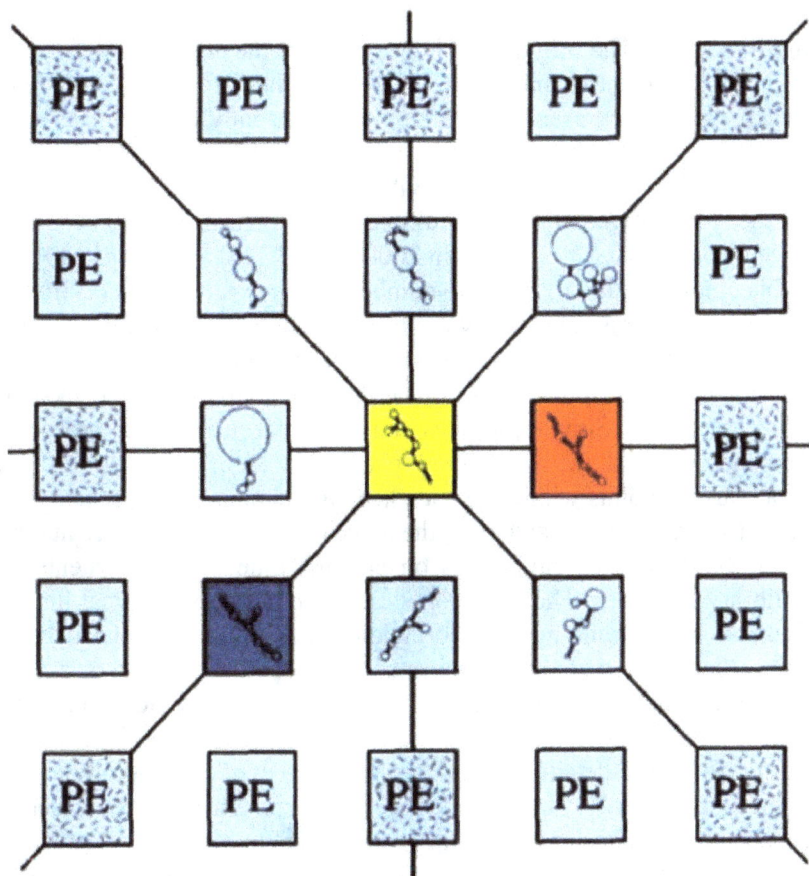

Figure 1. Illustration of the eight-way toroidally-wrapped connected grid of population elements (PEs) used to control the massively parallel Genetic Algorithm. Each element communicates with its nearest neighbors and evolves an RNA secondary structure in parallel. This representation is used on SIMD or MIMD architectures. The red and blue PEs represent the elements chosen as parents. The yellow represents the PE that will contain the newly generated chosen child structure for this neighborhood.

same holds true for the eastern and western "edges"). The location of an edge is therefore arbitrary, and has no meaning to the population itself, which is continuous. The massive parallelism of the algorithm refers to the fact that this population contains thousands or hundreds of thousands of elements, all evolving in parallel. The original implementation of the algorithm existed for the Single Instruction Multiple Data (SIMD) architecture MasPar MP2, on which each PE was represented by a single concurrently operating physical processor [Shapiro and Navetta, 1994]. Population sizes were limited then to a 16K (16,384) maximum. The current implementation of the algorithm utilizes the Multiple Instruction Multiple Data (MIMD) architecture of systems such as the CRAY/SGI Origin 2000 and T3E [Shapiro *et al.*, 2001a]. Under this implementation, a small number of physical processors run in parallel, each operating sequentially on its own subset of the population. The PEs in this case thus are virtual processors. The full logical layout described above is implemented by interprocessor communication via shared memory (shmem) library calls. Simple formulas convert between (physical processor, virtual processor) and (x, y) ordered pairs, and shmem synchronization barriers at various points within the code keep the population consistent from generation to generation. In actuality, the algorithm could run on any number (power of 2) of physical processors on any machine that implements the shmem library functions. Thus, the population size has no theoretical limit. It is difficult to define running time formulaically for this type of stochastic algorithm, especially when sequence-specific properties can have a large impact on the dynamic environment within the population. However, some empirical results can give an idea of the scalability of the algorithm and its population. (For more detailed results, see [Shapiro *et al.*, 2001a]). When the size of the population is varied while the number of physical processors is kept constant, the algorithm's running time appears to vary almost precisely linearly with respect to the ratio of virtual processors per physical processor. When the workload is varied by keeping population size constant while varying the number of physical processors, however, slight non-linearities arise due to interprocessor communication. A typical RNA sequence may be subjected to, say, twenty runs at a given population size, but we have found that significant information can be generated by comparing results at various population sizes as well, as described later.

Basic Algorithm

As the algorithm begins, a pool is generated containing all possible stems given a particular sequence. These stems may be either "fully zippered" or peeled back by a few base pairs. In addition, a secondary pool of "pseudostems" is generated, containing multiple-stem motifs that would be likely to occur simultaneously during the folding of a real molecule. Pseudostems are important in situations where a large rearrangement of the molecule must take place. If several correlated events must occur to facilitate this transformation, the probability of having all of the required stems present at one time may be too low to allow the transition. In current simulations, these pseudostem motifs have been limited to pairs of stems flanking a small symmetric internal loop across which coaxial stacking of helices may occur in reality, facilitating stem chain growth in natural systems. Both pools may be augmented by user-defined stems and motifs, or edited to delete particular interactions. On population initialization, each PE is randomly filled from these pools to generate diverse, random, and typically sparse structures throughout the population.

Once the population has been initialized, the iterated algorithm proper begins at generation 1. At each generation, the selection operator chooses and stores two parent structures for each PE location. Each PE generates a list containing itself and its eight nearest neighbors in the two-dimensional population layout. Conversion formulas are used to locate the virtual processor addresses corresponding to the neighbors in each direction from the central PE. This generated list then is ordered with respect to fitness, with the most thermodynamically stable structure in the highest position. A ranked rule biases probabilistic selection toward the head of the list for each parent. After the selection of parents, each PE generates two empty child structures, and mutations insert random stems into these structures from the stem pool. Longer sequences (with larger stem pools) are given higher mutation rates. Note that a pseudostem's component stems are treated independently except at the time of selection from the stem pool. If a stem chosen for mutation into a growing structure conflicts with a stem already in that structure, a probability exists that it will not be discarded, and will instead be peeled back to fit into the structure. This mechanism of peelback proved quite necessary to many simulations within the GA. The GA's basic unit of operation is a stem, as the algorithm inserts and deletes entire helical regions, but various natural structures contain helical regions that are much less than fully "zippered." This conflict-driven peelback mechanism, however, gives the GA the ability to increase its resolution to the level of single base-pairs of difference between structures. Increasing resolution at the time of stem insertion prevents the need to flood the stem pool itself with the non-relevant "noise" that would be introduced by including every possible stem at every possible degree of peelback. The mutation rate itself is not constant for the duration of the run. Instead, it follows an annealing curve that decreases the rate of mutation within each PE as the number of stems in that PE's structure grows [Shapiro and Wu, 1996]. Thus, the distribution of energy values across all PE's will gradually converge and the population will reach a consensus structure. A score measured by calculating the weighted standard deviation, i.e., over a limited window of past generations, of the average population-wide energy is used to determine the stop-point for the algorithm. The recombination event is implemented via a uniform crossover operation. First, after mutation is complete, one parent randomly distributes its stems into both child structures. Then, the second parent attempts to place its stems into both child structures. If at any point, a stem being added to one of the children conflicts with a previously inserted stem, there is again a probability that the conflict-peelback described above will occur. After each member of the population has completed the recombination phase, the structure in each PE is replaced by the better child structure. The selection, mutation, and crossover operations occur repeatedly until the score described above drops below a specified value, at which point the algorithm halts and reports a solution structure.

Additional Features

A number of additional features augment this basic behavior of the GA. H-type pseudoknots, for example, may be incorporated into structures with a minimal performance reduction of the algorithm. Pseudoknot stems are simply maintained in a separate list from the primary structure stems, and both lists are consulted for crossover operations and energy calculations. An H-type pseudoknot consists of a hairpin loop that pairs directly to the bases at the foot of its own stem and satisfies specific constraints [Shapiro and Wu, 1997]. Thus, either stem in the pair can be in either the pseudoknot list or the primary stem list, as either stem is valid independently. The GA also can simulate the effects of folding during transcription, or sequential folding. In this mode, the GA simply restricts the valid stem pool to those stems that fit within the sequence length at a given generation. The sequence is lengthened gradually, generation by generation at a rate defined by adjustable parameters. Also, a "Boltzmann Filter" may be activated which enhances the Boltzmann-like characteristics of the population [Wu and Shapiro, 1999]. Additionally, the algorithm may consider the stabilization of particular interactions by external factors not accounted for in the energy rules. If experimental evidence indicates that protein or ion stabilization, for example, affects a certain stem, that stem can be labeled within the algorithm as a "sticky stem." If a parent happens to contain this stem once selected, the parent is guaranteed to pass that stem on to both of its children. Thus, once one of these stems occurs naturally within the population, it can be encouraged to "stick," if it is favorable enough. Another method of more closely modeling natural conditions within the GA comes with the algorithm's multi-phase features. In natural systems, a molecule may pass through several phases during its lifetime, existing under different conditions in each. Within the algorithm, a solution structure from one phase can be seeded into a percentage of PE's during the initialization of a following phase, perhaps to be subjected to processing reactions or some other change of conditions. When running the GA in this manner, the non-seeded portion of the population is filled to a higher degree during the initialization phase, so that the random structures have a chance of competing with and providing alternatives for the relatively stable seeded structures. For example, one or two attempts may be made to insert stems into random structures during a regular run, while 2000 attempts may be made on a phase 2 run. Of course, many of the attempted insertions cannot be added due to conflicts, but keeping the number this high ensures that the random structures are as filled and stable as possible.

1.3 Data Generation and Analysis

Exemplary Biological Systems

Two biological systems happen to serve as particularly illustrative example cases for these methods. Potato Spindle Tuber Viroid (PSTVd) is a type-B viroid, one of several small, circular, unencapsidated RNA molecules that code for no proteins and infect a variety of plants, depending on host enzymes for replication. For reviews, see [Gross *et al.*, 1978; Diener, 1979a; Diener, 1979b; Keese *et al.*, 1988; Riesner *et al.*, 1990; Flores *et al.*, 1998; Diener, 1999]. Monomeric PSTVd has long been known to be able to form two very different structures *in vitro*. One of these is the native state, a highly stable, base-paired rodlike conformation (Figure 2a) [Riesner, 1979]. The other is a branched metastable state, containing three separate unusually stable regions: HPI, HPII, and HPIII (Figure 2b). The full rolling circle replication cycle of PSTVd (Figure 3) [Branch and Robertson, 1984;

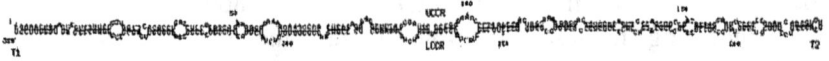

(a)

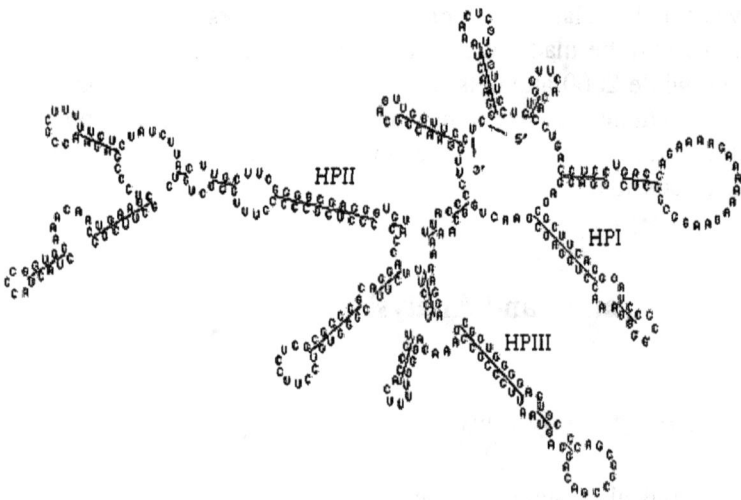

(b)

Figure 2. Depiction of the RNA secondary structures representing the fully folded monomeric PSTVd viroid 2a, and the metastable monomeric structure 2b. Both of these structures were predicted by the GA and have biological supporting evidence for their existence. The so-called HPI, HPII and HPIII unusually stable regions are also shown.

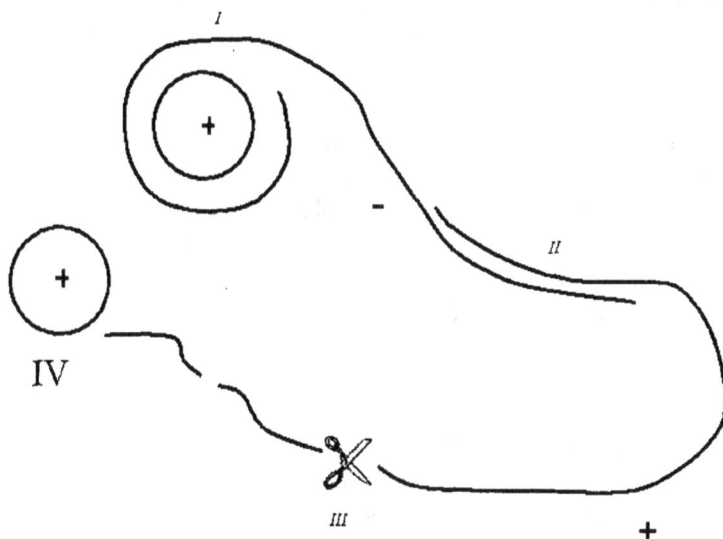

Figure 3. Depiction of the rolling circle model for the PSTVd life cycle. In stage I multimeric negative strand copies of the plus strand monmeric circular structure are generated. In stage II multimeric plus strands are produced. This is followed by stage III cleavage into monomeric linear structures which then form the plus strand monomeric circles.

Ishikawa *et al.*, 1984] involves the transcription of this circular monomer (+ strand) to a linear, multimeric (- strand) template. This template is then transcribed to multimeric (+ strand) copies, which are subsequently cleaved and ligated to monomeric circles. Experiments with both monomeric and dimeric clones of this molecule have demonstrated the potential for a number of structural transitions. The function of the metastable structures has been speculated upon at length, but the structure shown to be an active substrate for the cleavage reaction more closely resembles the rod-like conformation, with the addition of some unstable short-range interactions, which are stabilized by host factors.

The host-killing/suppression-of-killing (hok/sok) mechanism of *E. Coli* is a fairly complex system, meant to maintain plasmid copy number [Gerdes *et al.*, 1997]. The hok gene itself codes for a protein that will kill the cell if synthesized. In order for translation to occur, the hok mRNA must fold into a specific active conformation (Figure 4), including a translational activating interaction, and proper positioning of the Shine-Dalgarno elements of the two proteins for which the message codes, i.e., hok, and its regulator, mok (modulation of killing) [Franch and Gerdes, 1996]. This active conformation also contains a specific substructure that serves as a target for sok (suppression of killing) molecules. The plasmid that codes for hok and mok also codes for sok, a small antisense RNA. If the plasmid is present in sufficient copy number, sok is transcribed, and it binds to hok, causing degradation by RNase III and saving the cell [Gerdes *et al.*, 1992]. If the plasmid is not present in sufficient numbers, the killer protein is translated, and the faulty cell is prevented from replicating. Mechanisms must be in place, however, to prevent degradation of the pool of hok mRNA before replication; therefore these mechanisms must prevent binding of sok. This requires, however, that additional mechanisms prevent premature translation of the killer protein. The problem is compounded by the fact that hok mRNA folds during transcription, and thus the highly stable translational activating interaction can form before the antisense target region is even synthesized. Various studies have determined that, during transcription, the molecule is likely held in an inactive state (both non-translatable and non-sok-binding) by a metastable structure at its 5' end (Figure 5) [Gultyaev *et al.*, 1997; Nagel *et al.*, 1999]. Upon sequence completion, a long-range 5'-3' interaction locks the molecule into an inactive state (Figure 6) [Franch and Gerdes, 1996]. An active pool is slowly and continuously generated, however, by exonucleolytic processing at the 3' end [Franch *et al.*, 1997]. This processing truncates 30-40 nucleotides from the molecule, destroying the 5'-3' interaction and triggering a molecular rearrangement into the active state (both translatable and sok-binding) (Figure 4).

Population Dynamics and Interactive Visualization

While the GA is meant to study a dynamic, complex system, it happens to be quite such a system itself. Within the GA, there exists a lively arena of structural competition, statistical trends, and unexpected turns as the population explores the landscape defined by the particular RNA sequence in

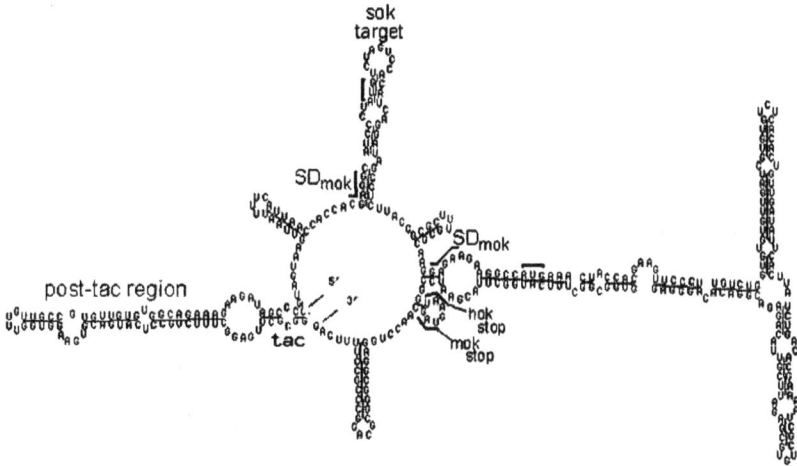

Figure 4. Representation of the secondary structure of the "active" conformation for the hok/sok plasmid RNA. This structure is produced during removal of several bases from the "inactive" structure's 3' end (see Figure 6) causing refolding of the molecule. Structures of particular importance are labeled. These include the sok target region where a small RNA will bind causing digestion of the RNA if the plasmid copy number is high enough. Also visible are the Shine-Dalgarno regions which have to be in their proper conformations for the hok killer protein to be synthesized if the plasmid copy number is too low.

use. It is important to find ways in which to monitor the progress of the population, and to understand its behavior in order to extrapolate from this environment into the reality of RNA folding pathways. The stochastic nature of the algorithm allows the PE's to move both uphill and downhill on the structure/energy landscape. The limited (nearest-neighbor) communication range allows different decisions to be made in various regions of the population, letting different groups of PE's explore different portions of this landscape and to discover information that might be passed over otherwise. Thus, at any instant during a particular run, there will be one or more distinct subpopulations present within the population as a whole. Each subpopulation consists of one or more contiguous regions with similar structures in each of

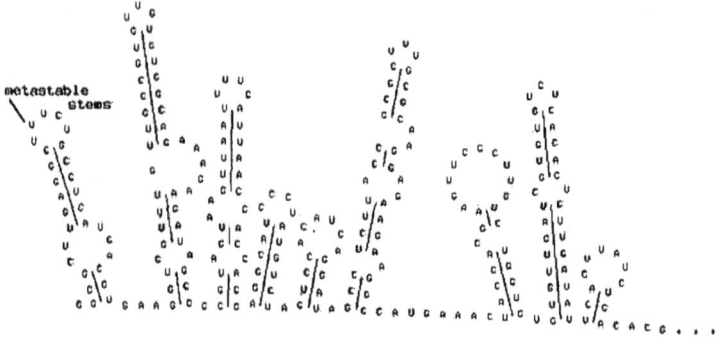

Figure 5. The early stages of the folding process of the hok/sok RNA. The temporary enforcement of the metastable stems shown here permit the structure to remain in an inactive state while the molecule is being synthesized to its full size. This is necessary to ensure that both premature degradation and premature synthesis of the killer protein are inhibited.

its PE's. There are various methods available to analyze the content and behavior of these subpopulations. The most direct view, however, giving at least a qualitative sense of algorithmic dynamics, is simply to look at them. The GA population may be visualized graphically in real time via several color-coded maps and associated histograms. These maps present displays of relative fitness, stem presence, and pseudoknot presence, and can provide information on the contents of a given PE at the click of a mouse. Figure 7 shows a snapshot of a fitness map of the population at generation 191 for a simulation of monomeric PSTVd at a population size of 16K. Although the molecule does not tend to fold as a monomer during the natural replication cycle, much experimental data is available for this particular molecular type, and the monomer provides for more elegant demonstration of some methods than the larger dimeric molecule. Please see [Shapiro *et al.*, 2001c] for a more detailed discussion of the dimeric PSTVd folding process.

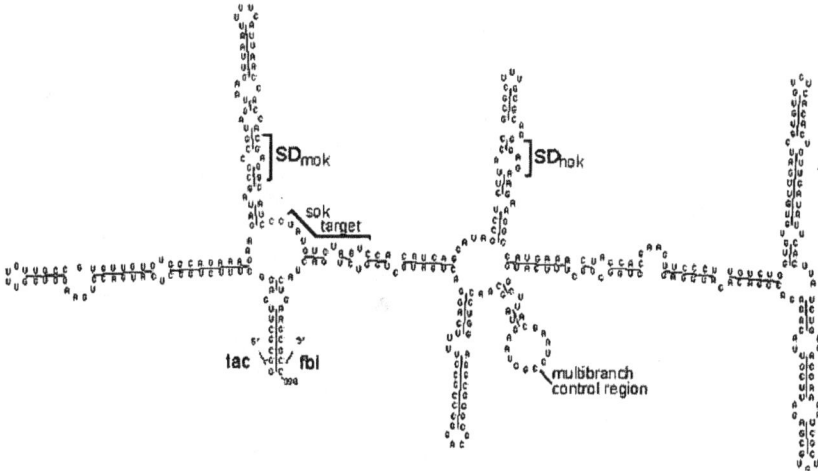

Figure 6. Representation of the secondary structure of the "inactive" conformation for the hok/sok plasmid RNA. This structure is produced upon completion of the synthesis of the entire RNA. Structures of particular importance are labeled. These include the sequestered Shine-Dalgarno regions of the mok and hok proteins as well as the absence of the sok binding region. A base paired region that helps retain this inactive conformation forms between tac and fbi.

   As the evolution process continues, subpopulations will expand and contract, covering various ranges within the 2-D layout. In the particular generation shown in Figure 7, a shrinking island containing a branched metastable structure is visible as the subpopulation of lowest fitness (area A). This subpopulation once was dominant during this run, and, as additional evidence will demonstrate, can be identified as the GA's solution structure for the PSTVd metastable state (see Figure 2b and the structure containing the three stems HPI, HPII, and HPIII in Figure 8C). The current dominant subpopulation is an intermediate structure generated by the GA. It is rodlike in the left-hand region, but contains a cruciform structure in the right-hand, or T2 domain (see Figure 7, area B, and Figure 8E). A small nucleation of PE's with yet lower energy (higher fitness) can be seen forming in the upper right hand corner (Figure 7, area C). The structure in this subpopulation is the

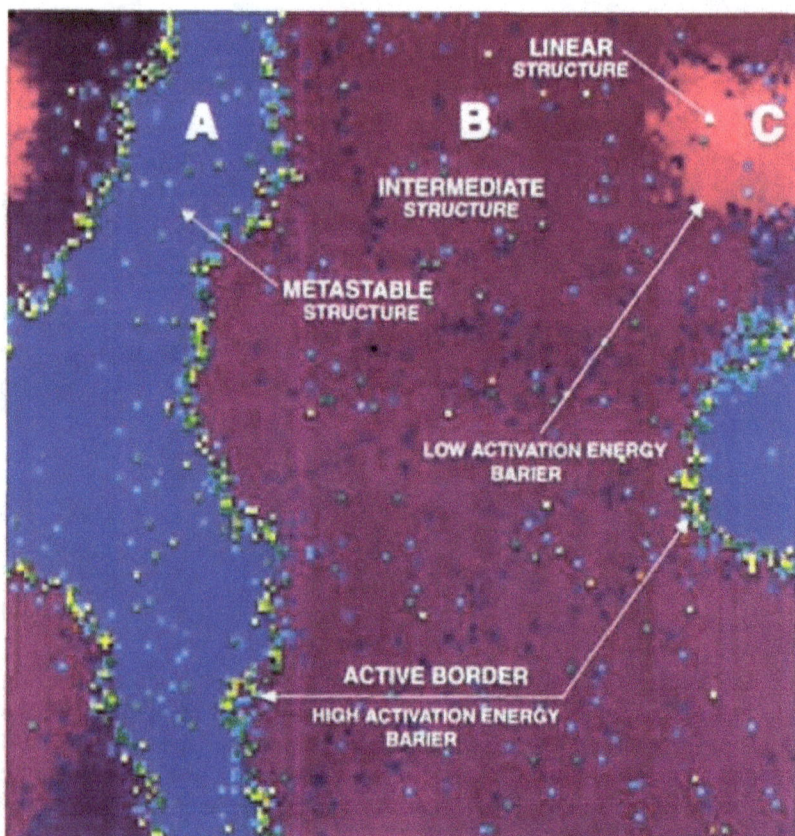

Figure 7. Fitness map depicting 16K population elements from generation 191 of the GA running the monomeric form of PSTVd. Each color coded pixel represents a fitness value of a particular structure. In this case blue represents poorer fitness than purple. The regions representing different structures are labeled in the image and are described in detail in the text.

native state, and will eventually subsume the rest of the population. When two structurally different subpopulations interact, the PE's in the border between the two must decide between two states. A transition from a less-stable local minimum to a more stable local minimum requires partial unfolding to accommodate new interactions. In a natural system, this would

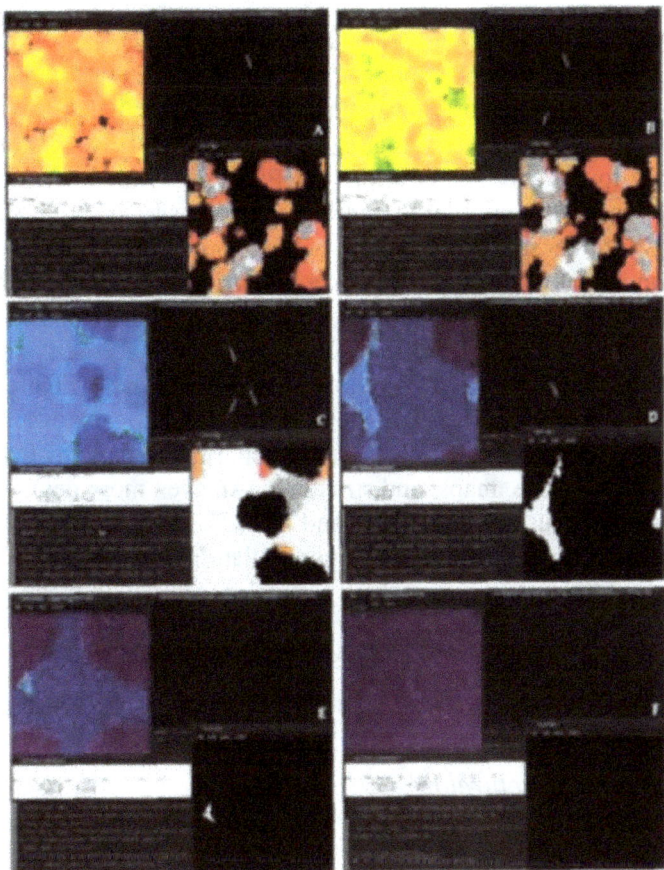

Figure 8. Selected frames, ordered left to right, top to bottom, from a GA run of the PSTVd monomer from the structures generated in a particular PE. Seen are the changing fitness maps for the entire population (in the upper left corner of each frame—red meaning poor fitness, purple good fitness); the changing depiction of a current structure in the chosen PE (in the upper right corner of each frame); and a trace map for the three important stems, in the metastable structure, that form the regions HPI, HPII and HPIII (shown in the lower right corner of each frame). These stems show their isolated existence, in the trace map, by the depiction of a particular color. The existence of more than one of these stems in a PE is indicated by a grayscale. The existence of all three stems in a PE is depicted by white. The absence of these stems is depicted by black. These stems are also highlighted in the structure drawings shown.

constitute the crossing of an activation energy barrier on the landscape. Within the GA, just such an event can be seen in the interaction between the metastable state and the cruciform state (Figure 7, border between areas A and B). In the "active border" region between the two subpopulations, a region of lower fitness develops as PE's fill with open intermediate structures (Figure 8D) in order to undergo transitions to the more stable state. The energy of the structures in this active border region can be used to estimate the height of the activation energy barrier. Note that the native state subpopulation formed spontaneously within the cruciform population. These two states are much more similar to one another, and the activation energy barrier is thus much lower, as is visible in the fitness map (Figure 7, border between areas B and C). Figure 8 shows six selected frames of a GA run of the monomer, showing in somewhat more detail the transitions that occur in the fitness maps, trace maps and structures. An even more detailed analysis of such transitions, using Stem Trace and STRUCTURELAB, is described in later sections. At a first-pass level, subpopulations may be correlated with structures by tracing the presence of particular stems within the population either numerically, or visually, using a map similar to the fitness map, but color-coded based on a list of trace stems provided by the user rather than on energy values.

Stem Trace and Structurelab

Stem Trace [Kasprzak and Shapiro, 1999] and the rest of STRUCTURELAB [Shapiro and Kasprzak, 1996] have proven indispensable in making sense of the complex data sets generated by the GA. Stem Trace itself is an abstract graphical plot of an ensemble of RNA (or DNA) structures, allowing the user to compare the stems present in a large number of structures quickly and informatively. Essentially, Stem Trace builds a plot as follows: as each new structure in an ensemble is inserted into the plot, the structure is assigned a coordinate along the x-axis. Each stem in that structure, uniquely defined by its 5' start position, 3' stop position, and length (and optionally by its energy), is assigned a position along the y-axis. If the stem has been plotted before in a previous structure, it is assigned the same y-coordinate as that of the first occurrence. Otherwise, it is assigned the next available (integral) y-coordinate. Thus, if a stem appears repeatedly within a plotted ensemble of structures, a horizontal band will appear at that y-coordinate. All color coding of Stem Trace plots presented in this text is by frequency of stem occurrence (Figure 9).

Stem Trace plots may be built from a series of structures consecutively reported as intermediates within the GA, from ensembles of structures representing the final solutions from multiple runs of the GA, from a set of suboptimal DPA solutions, and even from multiple sequences from the same family, using sequence alignments to correlate stems from each sequence. Depending on the situation, the ordering of stems along the y-axis may vary, including, but not limited to, order of appearance, 5' position, and energy. Any Stem Trace plot may be used to generate the associated, complementary, Single Strand Trace. In addition, Stem Trace functions as a graphical user interface to the data underlying the plot, and to other elements of STRUCTURELAB.

STRUCTURELAB as a whole provides a central, integrated interface, allowing access from a single workstation to a variety of tools implemented on various platforms and running concurrently. Drawing modules allow 2-D and 3-D visualization of structures, with automatic and interactive untangling and labeling features. Taxonomy tree plots may be used to compare an ensemble of structures, as may an interactive version of the familiar 2-D dot plots. Sequence and structural motif analysis may be carried out from within STRUCTURELAB, as may execution of the GA. For a more complete list of STRUCTURELAB features, see [Shapiro and Kasprzak, 1996].

Single Run Analysis

When one considers a single run of the GA, Stem Trace has the advantage of allowing one to take the population data from a potentially lengthy series of generations, and compile that data into a single plot. Detailed structural analysis of the simulated folding pathway then becomes a possibility. Since each x-coordinate in a Stem Trace plot represents one structure, the intuitive way in which to build a single run plot is to select one representative structure from the population at each generation. A variety of choices present themselves at this stage. Using the visualizer, a specific PE can be selected as the output point, based on criteria such as its interaction with particular subpopulations. Each structure that exists within this PE may then be plotted as a stem trace (see Figure 9, single processor data plot). This

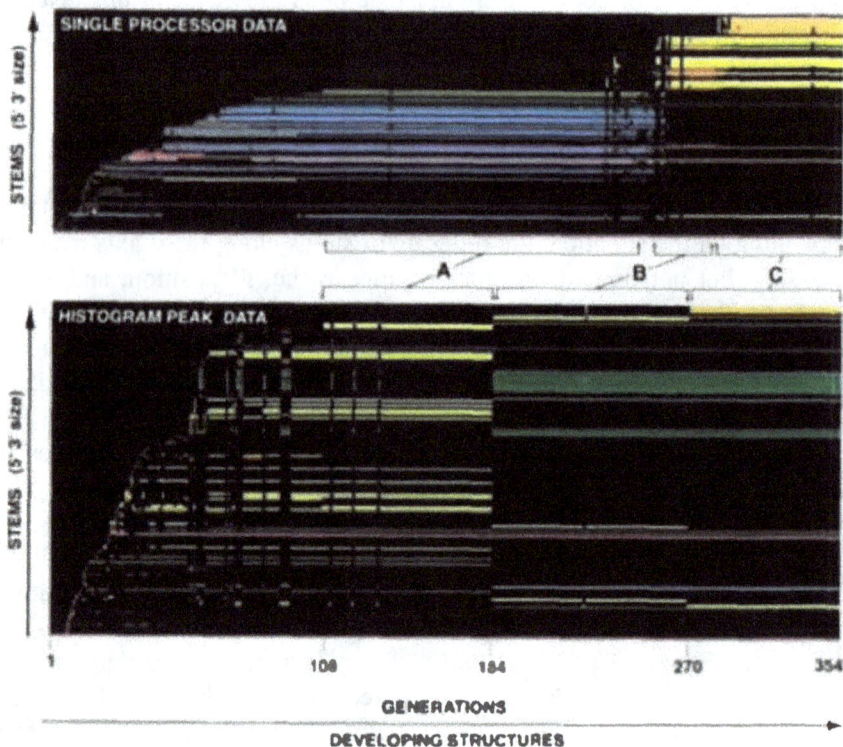

Figure 9. Depiction of two Stem Trace Plots and their correspondences. The top plot traces the progression of structural changes that take place in the PSTVd monomer, in a particular PE, over the course of a run of the GA. The generations that indicate the formation of three basic monomeric structures, A, B, C are shown. The bottom Stem Trace plot essentially shows the same transitions, but is based on the consensus structures of the entire population rather than the structures produced in a particular PE.

type of plot will provide a quantitative sense of the height of the activation energy barriers crossed by this PE as active borders sweep over its location. The particular structures plotted during these transitions, however, are difficult to analyze for significance, since they do not necessarily represent the behavior of the population as a whole. A second option, then, is to utilize the histogram of fitness values developed for the population at each generation. Selecting the peak in this histogram provides the most frequent energy value within the population. Selecting a structure with this specific energy value thus provides the majority structure within the population at that generation. A histogram peak stem trace therefore effectively plots the development of the population's consensus structure. Consider, for example, the histogram peak stem trace shown in Figure 9. This plot was compiled for the same monomeric PSTVd run depicted in Figures 7 and 8. There are clearly three distinct stages present, each represented by a different set of stems. These represent stages in the run during which three different subpopulations were dominant. The subpopulations labeled as A, B, and C in the time domain in this stem trace correspond directly to the subpopulations with the same labels in the space domain in the fitness map in Figure 7. Structures in population A are the metastable structures, containing the three metastable regions (Figure 2b, and the metastable-like structure in Figure 8C), structures in population B are the cruciform intermediates (Figure 8E), and structures in population C are the native rod (Figure 2a and Figure 8F). Note the intermittent appearance of stem bands from structure B within the structure A timeframe. This indicates that there was stiff dynamic competition between two similarly sized, alternative subpopulations for a time, but that structure A gained greater dominance until structure B evolved to a high enough fitness to compete more fully. The GA's results are consistent with experimental results demonstrating the ability of the branched metastable structure (defined experimentally by the presence of the three regions) to undergo a transition to the native rod (easily captured as an equilibrium structure). In addition, however, the GA is able to provide a complete structure for the metastable state, beyond the three main regions, and also to predict an intermediate structure, providing a detailed look at the molecule "in action." (Figure 8).

Multiple Run Analysis

A single run of the GA provides a compressed view of the series of stages that leads to a single solution, locating statistically significant structures from thousands of PE's over hundreds of generations. However, the stochastic nature of the algorithm means that, like the structures from a single processor, the solution from a single run must be compared to a larger set of data to determine its statistical significance. The solutions from multiple runs of the GA are typically compiled into a single Stem Trace plot in order to achieve this goal. A plot of this type puts each solution structure, and its component elements, into context, allowing the user to draw conclusions about what solutions are the GA's actual predictions, and what the significance of alternative structures may be. It is at this point that further compression of data, and therefore further determination of significant trends, can be performed. The mechanism for this layer of abstraction from the high level of complexity of the root level algorithm comes with the effects of population variation.

Multiple Population Size Analysis

As described in [Shapiro *et al.*, 2001a, 2001b and 2001c], variation of the size of the population operated on by the GA affects more than just the running time of the algorithm. The reason is as follows. Certain structural rearrangements are triggered by certain events within the population. In this case, an event essentially refers to the appearance of "the right stem in the right place at the right time." The smaller the population is, the more likely it is that the population as a whole will converge before a given event takes place. This situation is particularly noticeable when the energy/structure landscape for a particular sequence contains highly stable local minima with high activation energy barriers (i.e., metastable states). Subpopulations containing stable structures of this type can spread very rapidly throughout the population. If the population is small enough, this subpopulation can subsume the entire population before a more stable structure, which may occur at a later stage in the folding pathway, has the chance to arise and compete. The population becomes "kinetically trapped" behind an activation energy barrier. Certain population sizes may be particularly amenable to capturing specific states. As the population size is increased toward one of these values, the GA more and more deterministically produces the associated state as a solution. In this case, determinism is measured by comparing various values that measure the diversity within a solution ensemble for ensembles at different population sizes. Determinism can be estimated visually by comparing Stem Trace plots of these solution ensembles. A more deterministic set of runs has less variation in the ensemble, so it has a shorter Stem Trace plot with more high frequency stems. Note the 4K population size stem trace for monomeric PSTVd in Figure 10. The majority of structures at this population size are the metastable structure (Figure 2b), identical to that in areas labeled A in

Figures 7 and 9. As population size is increased the GA can begin to surmount the activation energy barrier, and can explore more of the landscape, as it moves toward another peak in determinism. This peak theoretically represents another, deeper local minimum, that is, a later stage in the folding pathway (note the 128K stem trace in Figure 10). The GA at this population size has located the rodlike native state (structure in Figure 2a, present in area C in Figures 7 and 9).

The practical upshot of this phenomenon is that smaller population sizes capture earlier states in the folding pathway and larger population sizes capture later states. Therefore, comparing solution structures as population is increased will generate a "consensus pathway." A Stem Trace of histogram peak structures over a single run compresses the data from many generations into a single plot. Similarly, the Stem Trace of solution structures over increasing population sizes compresses the data from many runs into a single plot. A population varying series of runs can determine which states are most significant to the pathway, and the order in which they occur. A single run can give a detailed picture of the full sequence of structures that successively gain dominance. And population visualization can provide the full-blown detail of how the structures in each PE are transforming. Each level of abstraction from the algorithmic processes gives hints about what is happening at a deeper level of complexity, and about how significant those happenings are to the whole story. The full range of analysis methods must be combined, and interpreted in the context of experimental data, to generate a complete picture of the folding pathway.

1.4 Biological Application

Below, we present examples of the application of these methods to multi-phase simulation of hok/sok and dimeric PSTVd. For the complete details on these results, please consult [Shapiro *et al.*, 2001c].

PSTVd

PSTVd has been long the subject of numerous experiments, quite understandably, since it has the potential to form several very different structures, and can shift between these states under the appropriate

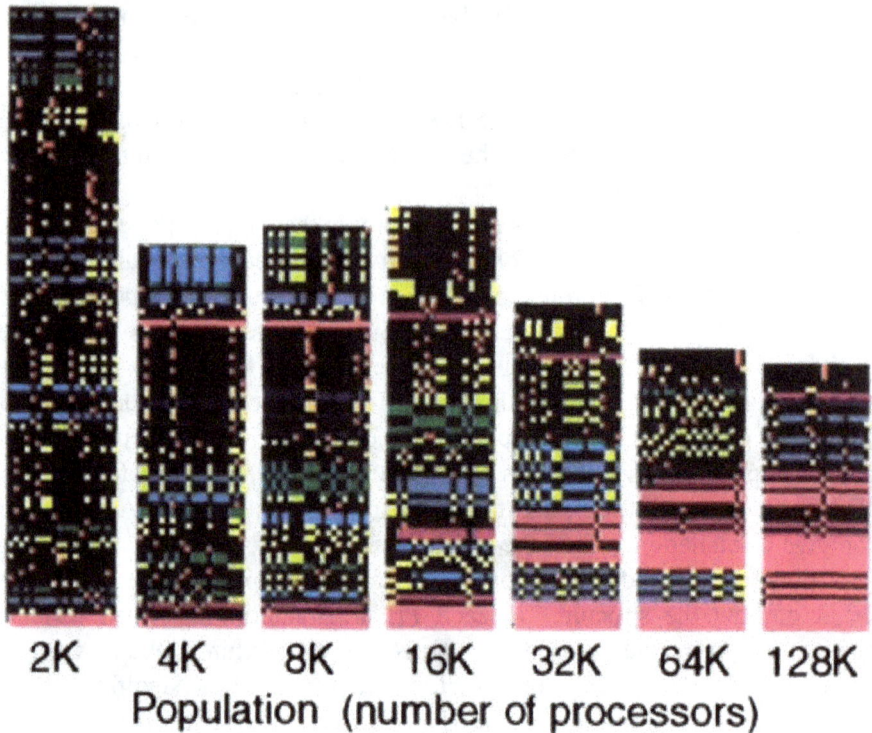

2K   4K   8K   16K   32K   64K   128K
Population (number of processors)

Figure 10. Depiction of Stem Trace results for the PSTVd monomer for varying populations sizes, i.e. 2K, 4K, 8K, 16K, 32K, 64K and 128K. Each population column contains the end results of 20 runs of the GA. The height of each column reflects the relative diversity of the structures that evolved in each population. The shorter the column the more deterministic the results. The levels of determinism are also reflected in the persistence (purple bands) of individual stems. The variation in the height of the columns is also reflecting the algorithm surmounting activation barriers (see text).

conditions. Earlier experiments in the absence of cellular extracts showed dimeric PSTVd molecules to assume, at equilibrium, the so-called tri-helical structure [Hecker et al., 1988]. This structure consists of two rodlike units joined by a perpendicular set of three helices, including two copies of HPI, formed by a long, nearly palindromic sequence in the upper central conserved region (UCCR). GA simulations analyzed with Stem Trace predicted formation of the same structure. The simulated folding pathway first showed formation of the same metastable structure seen in monomeric simulations. This holds well with the experimental detection immediately after transcription of a branched structure whose characteristics would be consistent with the presence of the three metastable regions [Hecker *et al.*, 1988]. Following the formation of this structure, we saw the subunits zipper into the rodlike conformation, while the HPI-trihelical region formed joining them. The GA structure predicted for the metastable intermediate is consistent with enzymatic probing results published in [Gast *et al.*, 1998].

As mentioned above, however, the active structure required for cleavage to monomeric units is not this tri-helical structure. Instead, experimental evidence has shown that the active structure consists of a significantly suboptimal state [Steger *et al.*, 1992; Baumstark and Riesner, 1995; Baumstark *et al.*, 1997]. This state contains several hairpins in the UCCR, hairpins that have been shown to require nuclear extract proteins for stabilization [Tsagris *et al.*, 1987; Baumstark *et al.*, 1997]. This so-called Extended Middle structure first was discovered in experiments using a monomeric construct extended in the UCCR with the minimum portion of a second subunit that would still promote cleavage. This system is an ideal case for the application of sticky stems (see above). By marking as sticky stems the two hairpins for which both experimental and phylogenetic evidence existed, we were able to simulate external stabilization of these structures within the GA. Upon doing so, the GA indeed predicted formation of the Extended Middle structure, and even independently predicted formation of a third hairpin, one suggested in the literature, but for which experimental evidence had not been found.

The success of this simulation made it reasonable to include the sticky stems in the full dimeric simulation. Upon doing so, the GA predicted formation of a double Extended Middle structure instead of the tri-helical structure. The multi-phase capabilities of the GA could then be employed to explore the refolding events that would occur after cleavage at the experimentally known cleavage site. UV crosslinking experiments have shown that the molecule, once cleaved to a monomeric state, is driven into the correct conformation for ligation by the formation of a Loop-E type structure [Branch *et al.*, 1985; Baumstark *et al.*, 1997]. When seeding the Extended Middle result of phase 1 into a phase 2 run, and truncating the molecule at the known cleavage sites, we were able to explore the GA's version of this transition. In such secondary phase runs, it is generally profitable to build the single run Stem Traces by selecting the best (fittest) structure within the population at each generation, rather than the histogram peak structure. This procedure follows the refolding seeds rather than the noise of the highly randomly filled surrounding population described above. Interestingly, the first transition that appeared in this case included HPI, a structure whose importance to the replication cycle has been debated. HPI appeared able to stabilize quickly the large free-stranded region generated by the cleavage reaction. Subsequently, the GA indeed showed formation of the Loop-E structure. This structure, which theoretically contains non-canonical base pairs, is thought to be stabilized by Mg^{2+} ions [Gast *et al.*, 1996; Baumstark *et al.*, 1997]. In the GA simulation, a small hairpin stem formed in the upper-strand portion of Loop-E. This stem could have been an artifact of the simulation, but it was able to play the role of Mg^{2+} in stabilizing the large loop. In a third phase, the ligation reaction was simulated by shifting the 5'-3' gap from Loop-E to the left-hand, or T1 domain, which had already formed a closed stem-loop. Doing so caused the Loop-E hairpin to unfold and open the loop once the strand became continuous.

hok/sok

The host-killing/suppression-of-killing system proved to be quite interesting in the context of the GA. One useful property of this system was that it could act as a test case for the sequential folding features of the GA. While at appropriate population sizes, the GA would indeed find the correct, inactive structure for hok mRNA; closer examination of the folding pathway revealed that the molecule was passing through an active conformation in intermediate stages. The stems identified in [Gultyaev *et al.*, 1997; Nagel *et al.*, 1999] as inactivating metastable structures were indeed forming during sequential folding, and were present in nearly 100% of the population. But the more stable translational activating (tac) interaction would slowly replace them once its nucleotides had been synthesized. In order to assess the influence of the metastability of these stems on the folding pathway, we enforced it in all structures up to the point in the sequence synthesis at which these stems usually began to drop out of the histogram peak subpopulation. When enforcement was removed at this point, nothing prevented the more stable tac interaction from forming, yet it did not form. The structure remained in an inactive conformation until the 5'-3' interaction could lock in the inactive state. Regardless of the pathway, however, the final structure was always the same. That is, the molecule would arrive in the identical inactive state regardless of whether it did or did not pass through an intermediate active state. Thus, the GA demonstrates the importance of intermediate stage interactions, even when they have no influence on the native state equilibrium structure.

The central multibranch junction in this molecule was a rather interesting substructure in and of itself. In the first place, it contained an unusually large number of stems that had to be peeled back to quite a large degree in order to accommodate one another. In fact, it was this molecule that demonstrated the need for the conflict-driven peelback method within the GA. This method was required to reach the correct inactive structure within the simulation. In addition, the correct structure for this junction was by no means optimal. One of the branches, labeled as a control region, in Figure 6, was required to maintain this suboptimal, inactive structure. Without this branch, the molecule was approximately 3 kcal more stable, but the Shine-Dalgarno element of hok was no longer properly sequestered. When marking this control region branch with the sticky stem mechanism, we observed formation of the correct suboptimal structure. This raises the possibility that this structure is stabilized by something not accounted for by the basic energy rules.

A population-varying run identified a structure that was increasingly prevalent as population size was decreased from that which located the native state most deterministically. Instead of the long-range 5'-3' interaction, the 5' end of the molecule formed a very stable local interaction. This substructure tied up part of the translational activating interaction by pairing it with the downstream sequence normally forming the SD_{hok}-sequestering stem. Interestingly, we were able to locate similar interactions in the other members of the hok family.

Hok mRNA offers a clear condition for a second GA phase. For phase 2, we seeded the inactive native state structure into the population and slowly truncated the 3' end as the algorithm progressed, simulating the exonucleolytic 3' processing. Indeed, a global rearrangement was triggered that placed the molecule into a state identical in virtually every base-pair to the published active structure. Closer examination of the refolding pathway, however, revealed a striking intermediate stage. An alternative structure to the solution structure clearly formed an extremely self-consistent sub-ensemble, which increased in proportion to the entire solution ensemble size as GA population size was decreased. At even lower population sizes, similarly distinct structures appeared, representing yet earlier phases. The most notable intermediate substructure was a pairing between the 5' end of the molecule and nucleotides in the antisense target structure [Shapiro *et al.*, 2001c]. This interaction would not only prevent the translational activating interaction from forming, but it would directly prevent the antisense target structure from forming as well. Thus, unlike the other mechanisms of inactivation, this novel pairing provided a direct, steric block to both functions of the active structure. Moreover, phylogenetic analysis revealed that the interaction was extremely conserved across the entire hok family by significant numbers of compensatory base changes within the stem.

1.5 Conclusions

The massively parallel Genetic Algorithm seems to have great potential for the exploration of RNA folding pathways. When conditions for folding within the algorithm are adjusted to match those of experimental environments, the algorithm appears to report similar results to the original experiments. The strength of this approach is that, once one has correctly adjusted the computational system to match the biological system, the simulation can provide a wealth of information that would be difficult to gain from experiments alone. The algorithm offers the opportunity to catch detailed glimpses of intermediate structures that are challenging to capture and directly analyze experimentally. One could employ this capability either for verification or for prediction of the dynamic structural details of a molecule's behavior. In actuality, a combination of both approaches seems most useful. The most effective use of the algorithm is not in a vacuum, but as applied to a system about which there already exists some information. Algorithmic, experimental, and phylogenetic analyses can then mutually support and extend one another.

The methods described here illustrate the importance of having a variety of effective ways of visualizing the same data from many perspectives. The GA deals with a complex system and generates a large amount of data very rapidly. The various levels of abstraction and compression of this data into numerical and graphical representations are crucial for making sense of it all. Stem Trace has proven to provide some of the most valuable of these representations, and has been indispensable for the generation of these results.

Many hold the view that the folding process of RNA is hierarchical, that primary sequence first defines secondary structure, and that tertiary structure subsequently forms as a consequence of secondary structure. If this is the case, analyses of such systems as are described here can be carried out on each level independently, and maintain validity. However, full understanding of a system is only possible with the integration of analyses on all three levels. Thus, future directions with the GA should include an increase in its ability to consider the contributions of tertiary structure, as well as the analysis of GA-generated data by methods designed for tertiary structure analysis.

Acknowledgment

This publication has been funded in part with Federal funds from the National Cancer Institute, National Institutes of Health, under Contract No. N01-C0-12400.

References

Baumstark, T. and Riesner, D. (1995) "Only one of four possible secondary structures of the central conserved region of potato spindle tuber viroid is a substrate for processing in a potato nuclear extract." *Nucl Acids Res.* **23(21)**, 4246-4254.

Baumstark, T., Schröder, A.R.W. and Riesner, D. (1997) "Viroid processing: switch from cleavage to ligation is driven by a change from a tetraloop to a loop E conformation." *EMBO J.* **16(3)**, 599-610.

Branch, A.D., Benenfeld, B.J. and Robertson, H.D. (1985) "Ultraviolet light-induced crosslinking reveals a unique region of local tertiary structure in potato spindle tuber viroid and HeLa 5S RNA." *Proc. Natl. Acad. Sci.* **82(10)**, 6590-6594.

Branch, A.D. and Robertson, H.D. (1984) "A replication cycle for viroids and other small infectious RNA's." *Science* **223(4635)**, 450-455.

Diener, T.O. (1979a) "Viroids: structure and function." *Science* **205(4409)**, 859-866.

Diener, T.O. (1979b) *Viroids and Viroid Diseases.* John Wiley & Sons, Inc., New York.

Diener, T.O. (1999) "Viroids and the nature of viroid diseases." *Arch. Virol. Suppl.* **15**, 203-220.

Flores, R., Randles, J.W., Bar-Joseph, M. and Diener, T.O. (1998) "A proposed scheme for viroid classification and nomenclature." *Arch. Virol.* **14**(3), 623-630.

Franch, T. and Gerdes, K. (1996) "Programmed cell death in bacteria: translational control by mRNA end-pairing." *Molecular Microbiology* **21**(5), 1049-1060.

Franch, T., Gultyaev, A.P. and Gerdes, K. (1997) "Programmed cell death by hok/sok of plasmid R1: Processing at the hok mRNA 3' triggers structural rearrangements that allow translation and antisense RNA binding." *J. Mol. Biol.* **273**, 38-51.

Gast, F., Kempe, D. and Sänger, H.L. (1998) "The dimerization domain of potato spindle tuber viroid, a possible hallmark for infectious RNA." *Biochemistry* **37**, 14098-14107.

Gast, F., Kempe, D., Spieker, R.L. and Sänger, H.L. (1996) "Secondary structure probing of potato spindle tuber viroid (PSTVd) and sequence comparison with other small pathogenic RNA replicons provides evidence for central non-canonical base pairs, large A-rich loops, and a terminal branch." *J. Mol. Biol.* **262**, 652-670.

Gerdes, K., Gultyaev, A.P., Franch, T., Pedersen, K. and Mikkelsen, N.D. (1997) "Antisense RNA-regulated programmed cell death." *Annu. Rev. Genet.* **31**, 1-31.

Gerdes, K., Nielsen, A.K., Thursted, P. and Wagner, E.G.H. (1992) "Mechanism of killer gene reactivation: antisense RNA mediated RNaseIII cleavage ensures rapid turn-over of the hok, srnB, and pndA effector mRNAs." *J. Mol. Biol.* **226**, 637-649.

Gross, H.J., Domdey, H., Lossow, C., Jank, P., Raba, M., Alberty, H. and Sänger, H.L. (1978) "Nucleotide sequence and secondary structure of potato spindle tuber viroid." *Nature* **273**, 203-208.

Gultyaev, A.P., Franch, T. and Gerdes, K. (1997) "Programmed cell death by hok/sok of plasmid R1: Coupled nucleotide covariations reveal a phylogenetically conserved folding pathway in the hok family of mRNAs." *J. Mol. Biol.* **273**, 26-37.

Hecker, R., Wang, Z., Steger, G. and Riesner, D. (1988) "Analysis of RNA structures by temperature gradient gel electrophoresis: viroid replication and processing." *Gene* **72**, 59-74.

Ishikawa, M., Meshi, T., Takeshi, O., Yoshimi, O., Teruo, S., Ichiro, U. and Shikata, E. (1984) "A revised replication cycle for viroids: The role of longer than unit length RNA in viroid replication." *Mol. Gen. Genet.* **196(3)**, 421-428.

Kasprzak, W. and Shapiro, B.A. (1999) "Stem Trace: An interactive visual tool for comparative RNA structure analysis." *Bioinformatics* **15(1)**, 16-31.

Keese, P., Visvader, J.E. and Symons, R.H. (1988) "Sequence variability in plant viroid RNAs." In *RNA Genetics*, eds. Domingo, E., Holland, J. J. and Ahlquist, P., vol. 3, p. 71-98. CRC Press, Boca Raton, FL.

Mathews, D.H., Sabina, J., Zuker, M. and Turner, D.H. (1999) "Expanded sequence dependence of thermodynamic parameters improves prediction of RNA secondary structure." *J. Mol. Biol.* **288**, 911-940.

Nagel, J.H.A., Gultyaev, A.P., Gerdes, K. and Pleij, C.W.A. (1999) "Metastable structures and refolding kinetics in hok mRNA of plasmid R1." *RNA* **5**, 1408-1419.

Riesner, D. (1990) "Structure of viroids and their replication interediates. Are thermodynamic domains also functional domains?" *Virology* **1**, 83-99.

Riesner, D., Henco, K., Rokohl, U., Klotz, G., Kleinschmidt, A. K., Domdey, H., Jank, P., Gross, H.J. and Sänger, H.L. (1979) "Structure and structure formation of viroids." *J. Mol. Biol.* **133**, 85-115.

Rivas, E. and Eddy, S.R. (1999) "A dynamic programming algorithm for RNA structure prediction including pseudoknots." *J. Mol. Biol.* **285**, 2053-2068.

Shapiro, B.A., Bengali, D., Kasprzak, W. and Wu, J-C. (2001b) "Computational insights into RNA folding pathways: Getting from here to there." In *Proceedings of the Atlantic Symposium on Computational Biology and Genome Systems and Technology.*

Shapiro, B.A., Bengali, D., Kasprzak, W. and Wu, J-C. (2001c) "RNA folding pathway functional intermediates: Their prediction and analysis." *J. Mol. Biol.* **312(1)**, 27-44.

Shapiro, B.A. and Kasprzak, W. (1996) "STRUCTURELAB: A heterogeneous bioinformatics system for RNA structure analysis." *Journal of Molecular Graphics* **14**, 194-205.

Shapiro, B.A. and Navetta, J. (1994) "A massively parallel genetic algorithm for RNA secondary structure prediction." *The Journal of Supercomputing* **8**, 195-207.

Shapiro, B.A. and Wu, J-C. (1996) "An annealing mutation operator in the genetic algorithm for RNA folding." *CABIOS* **12**(3), 171-180.

Shapiro, B.A. and Wu, J-C. (1997) "Predicting RNA H-type pseudoknots with the massively parallel genetic algorithm." *Bioinformatics* **13**(4), 459-471.

Shapiro, B.A., Wu, J-C., Bengali, D. and Potts, M.J. (2001a) "The massively parallel genetic algorithm for RNA folding: MIMD implementation and population variation." *Bioinformatics* **17**(2), 137-148.

Steger, G., Baumstark, T., Mörchen, M., Tabler, M., Tsagris, M., Sänger, H.L. and Riesner, D. (1992) "Structural requirements for viroid processing by RNase T1." *J. Mol. Biol.* **227**, 719-737.

Tsagris, M., Tabler, M., Mülbach, H. and Sänger, H.L. (1987) "Linear oligomeric potato spindle tuber viroid RNAs are accurately processed in vitro to the monomeric circular viroid roper when incubated with a nuclear extract form healthy potato cells." *EMBO J.* **6**(8), 2173-2183.

Wu, J-C. and Shapiro, B.A. (1999) "A Boltzmann filter improves the prediction of RNA folding pathways in a massively parallel genetic algorithm." *Journal of Biomolecular Structure and Dynamics* **17**(3), 581-595.

Zuker, M. (1989) "On finding all suboptimal foldings of an RNA molecule." *Science* **244**, 48-52.

Authors' Addresses

Bruce A. Shapiro, Laboratory of Experimental and Computational Biology, The NCI Center for Cancer Research, NCI-Frederick, National Institutes of Health, Building 469, Room 150, Frederick, MD 21702, USA.
Email: bshapiro@ncifcrf.gov.

David Bengali, Laboratory of Experimental and Computational Biology, The NCI Center for Cancer Research, NCI-Frederick, National Institutes of Health, Building 469, Room 150, Frederick, MD 21702, USA.

Wojciech Kasprzak, Intramural Research Support Program, SAIC, NCI-Frederick, Frederick, MD 21702, USA.

Jin Chu Wu, Intramural Research Support Program, SAIC, NCI-Frederick, Frederick, MD 21702, USA.

Chapter 2

Introduction to Self-Assembling DNA Nanostructures for Computation and Nanofabrication

Thomas H. LaBean

2.1 Introduction

DNA, well-known as the predominant chemical for duplication and storage of genetic information in biology, has also recently been shown to be highly useful as an engineering material for construction of special purpose computers and micron-scale objects with nanometer-scale feature resolution. Properly designed synthetic DNA can be thought of as a programmable glue which, via specific hybridization of complementary sequences, will reliably self-organize to form desired structures and superstructures. Such engineered structures are inherently information-rich and are suitable for use directly as computers or as templates for imposing specific patterns on various other materials. In theory, DNA can be used to create any desired pattern in two or three dimensions and simultaneously to guide the assembly of a wide variety of other materials into any desired patterned structure. Given diverse mechanical, chemical, catalytic, and electronic properties of these specifically patterned materials, DNA self-assembly techniques hold great promise for bottom-up nanofabrication in a large number of potential applications in wide ranging fields of technology. Starting with background for understanding why the physical, chemical, and biological properties of DNA make it extremely useful as a "smart" material for nanoengineering projects, this chapter traces the historic development of DNA-based nanofabrication, outlines its major successes, and presents some possible future applications in fields as diverse as electronics, combinatorial chemistry, nano-robotics, and gene therapy.

DNA-based nanoengineering as a field is related to computational biology, bioinformatics, and genome informatics rather tangentially; it is more closely allied with biomolecular computation (BMC) – the engineering of biological macromolecules for production of artificial information processing systems. Rather than using binary, electronic computers for analyzing information extracted from biological systems, BMC seeks to utilize biomolecules directly as active parts of engineered computers. The concluding section of this chapter contains some speculation into the possibility of coming full circle and applying BMC and DNA-based nanoengineering principles and systems to the extraction and processing of information directly from biological DNA, that is, the possible use of natural DNA molecules as inputs for artificial DNA-based machines.

2.2 Background

Chemistry and Biology of DNA

DNA (deoxyribonucleic acid) is a linear polymer whose monomeric residues are made up of one sugar group (deoxyribose), one phosphate group, and one nitrogenous base (either adenine, cytosine, guanine, or thymine; designated A, C, G, and T, respectively). Details of the structure are available in any biochemistry or molecular biology textbook, but a few pertinent points will be mentioned here. First, neighboring residues are joined by a chemical bond between the nth phosphate and the $(n+1)^{th}$ sugar group such that a polymeric backbone is formed of alternating sugar and phosphate groups. The backbone has chemical directionality due to asymmetry in the placement of phosphate groups on the sugar, with each sugar having one phosphate bound to its 5' carbon and one phosphate bound to its 3' carbon. This asymmetry gives the entire polynucleotide chain two distinct ends – the 5' and the 3', as shown in Figure 1. Two DNA strands hybridize (form hydrogen bonds) to one another in anti-parallel fashion, thus the 5' end of one strand points toward the 3' end of its complementary strand in the famous Watson-Crick double-stranded form (or double helix).

The second pertinent point regarding the chemical nature of DNA is that the nitrogenous bases (or simply bases) form hydrogen bonded pairs in tongue-and-groove fashion providing specificity of annealing. The base groups decorate the sugar-phosphate backbone with regular spacings and provide the physico-chemical energy which zips the DNA together in its predictable helical structure. In double-helical DNA (or double-strand DNA, abbreviated to dsDNA), G bases pair specifically with C residues and A bases pair with T bases. G and C are said to be *complementary*, as are A and T. DNA strands of exact Watson-Crick complementarity will form stable hydrogen-bonded structures under standard temperature and solution conditions (see Figures 1 and 2). Some alternative base pairings have been found to form fairly stable hydrogen bonding (see for example [Peyret *et al.*, 1999]), however, careful design of the sequences, as well as very slow annealing protocols, can successfully avoid alternative pairings and ensure that perfectly complementary strand matchings are highly favored. The third important point stems directly from the exceptional stability and specificity of dsDNA. If a short segment of single-strand (ssDNA) is appended to a longer strand which participates in a double-helical domain, the ssDNA will act as a "smart glue", binding specifically to a complementary ssDNA segment located on another ds-domain. These ssDNA segments are known as *sticky-ends*. Complementary sticky-end pairs therefore act as address labels and can be used to specify which dsDNA domains are allowed to anneal to one another.

Finally, the "folding rules" which dictate the three-dimensional (3D) structure of DNA in solution are simple compared to other biological macromolecules, making DNA a more salutary engineering material than proteins, for example, whose folding rules have yet to be completely understood. Given proper pH and cation concentration, dsDNA will reliably adopt standard B-form helical structure with predictable dimensions as shown in Figure 1. In summary, important points of DNA chemistry include: anti-parallel alignment of backbones in hybridized strands, base-pairing specificity for high-fidelity annealing of sequences to their complements, and annealing by heating and slow cooling for double helix formation.

The task of engineering specific physical structures from DNA benefits

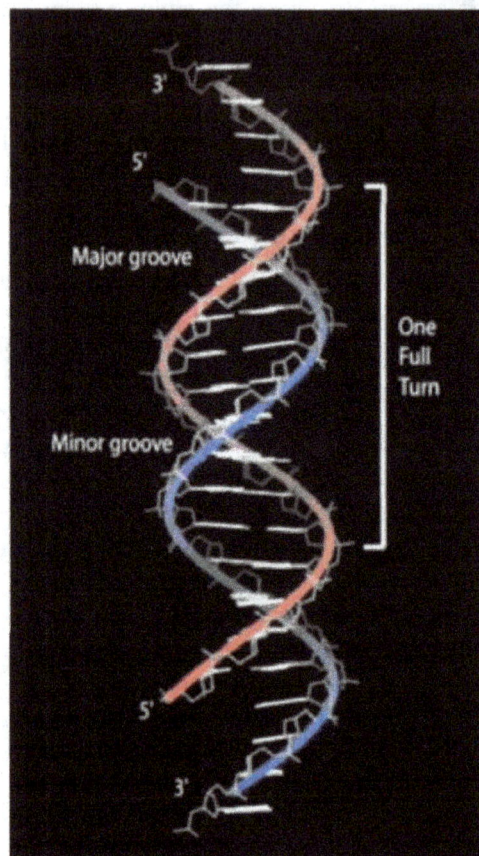

Figure 1. Double-stranded DNA shown in the standard, right-handed, B-form double helix with four base ssDNA sticky-ends appended to the 3' ends of both strands. Strand backbones are highlighted with colored ribbons; bases (light gray) are viewed edgewise and can be seen to point toward their hydrogen bonding partner on the opposite strand. One full turn of DNA has a length of 3.4 nm along the vertical helix axis and contains on average 10.5 bases; the helix diameter is approximately 2 nm. The concave faces of the helix are known as the major and minor grooves; they are geometrically distinct and can be used to identify strand polarity -- for example, when looking into the minor groove, the strand on the bottom (in this orientation) always has its 3' end pointing down (toward the bottom of the page). Understanding the geometric constraints of DNA structure is essential to successful design of DNA-based objects and materials.

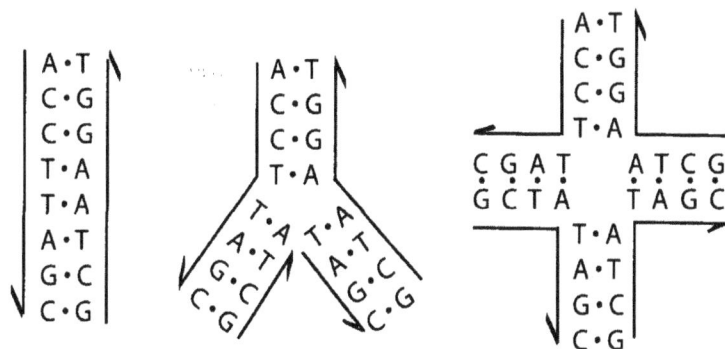

Figure 2. Representations of unbranched, 3-branched, and 4-branched DNA. Fully base-paired, anti-parallel DNA can take on various forms depending upon the lengths and connectivities of the annealed strands. Normal double-helical DNA (left) involves two strands in a single helical domain. A 3-branched junction (center) involves four strands and three helical domains; it is a structural analog of a replication fork observed in biology. The 4-branched junction (right), involving four strands and four helical domains, is a structural analog of the Holliday junction used by biological systems in genetic recombination. Note that, in the branched structures shown, alternative base-pairings are available due to sequence symmetry around the branch point which will allow the junction to migrate up and down the helices. Properly designed sequences avoid such migration. 4-branch junctions have been used most extensively in engineered tile structures. Their four helical domains tend to stack into two domains in which two strands exchange between helices (as explained further in the next figure).

from the tools evolved during the eons of biological evolution on Earth and especially from those now thoroughly researched and commercialized during the more recent biotechnological revolution. Enzymes can be purchased which perform highly specific chemical reactions upon DNA molecules. For example, *phosphatases* and *kinases* remove, add, and exchange phosphategroups from the ends of DNA backbones; *ligases* stitch together breaks in the backbone to form a single chemical strand from two or more shorter strands; and *restriction endonucleases* cleave the backbone at specific sites dictated by local base sequence. In addition, chemical synthesis methods for the production of DNA have advanced to the point where DNA strands of any desired sequence can be ordered on-line from commercial production companies and shipped the next day for less than a dollar per residue.

DNA as a Structural Material

Since the publication of the 3D structure of dsDNA half a century ago [Watson and Crick, 1953], the vast majority of research on DNA structure has centered around DNA as it relates to known biological systems. However, twenty years ago Nadrian Seeman recognized the inherent potential of DNA as an engineering material and proposed visionary new uses for the polymer [Seeman, 1982]. Seeman's pioneering work originally focused on the creation of regular 3D lattices of DNA which could be used as scaffolding for the rapid, orderly binding of proteins to speed the formation of suitable crystals for 3D protein structure elucidation in x-ray diffraction studies.

Seeman noted that linear dsDNA can interact with only two other double-helices since it can display at most two sticky-ends, i.e. its maximum valence is two. Construction materials with valence = 2 are only really useful for making linear superstructures like railroad cars connected in a long train. A larger variety of substructures and an ability to interact with a greater number of neighboring components is required in order to advance even modest fabrication goals. Seeman pointed out that DNA in biological systems can exhibit structures with increased valence including replication forks (valence = 3) and Holliday junctions found in genetic recombination (valence = 4) as shown in Figure 2. One problem with these natural multivalent structures is that they involve repeated base sequences, so base-pairing partners are not perfectly specified and the junctions are mobile. The junctions, or strand crossover points, between the dsDNA domains are free to migrate up or down the helices by swapping one perfect sequence match for another perfect sequence match (see the right-hand drawing in Figure 2, if the top helix is pulled up while the bottom helix is pulled down, the left and right helices will become shorter as the top and bottom helices become longer). Seeman worked out a sequence symmetry minimization strategy in order to form, for the first time, immobile junctions - - branch points in the dsDNA which are unable to migrate up and down the helix. Note that the oligonucleotides are still normal, linear DNA polymers; the branch junctions occur in the arrangement of strand exchange crossovers between the double helical arms. Seeman has pioneered the use of branched DNA structures for the construction of geometric objects, knots, and Borromean rings [Chen *et al.*, 1989; Chen and Seeman, 1991; Du and Seeman, 1994; Zhang and Seeman, 1994]. These early construction projects yielded many important technical developments including the use of oligonucleotide assemblies bound to insoluble resin beads for control of construction.

One problem with many early DNA constructs was that the structural flexibility of the branched DNA complexes allowed undesired circular products to be formed during assembly of large superstructures from stable substructures. Again, innovation from Seeman's lab solved the problem by producing double-crossover (DX) complexes [Fu and Seeman, 1993] which act as rigid structural components for assembly of larger superstructures. The concept has now been extended further to produce more complex structures including triple-crossovers (TX) [LaBean *et al.*, 2000b] as shown in Figure 3. This class of DNA objects, often referred to as 'tiles', contain multiple oligonucleotide strands (ssDNA) which base-pair along parallel, coplanar helix axes. The helices are connected by exchange of two strands at each crossover point (crossovers are structural analogs of Holliday junctions). Rigid and thermally stable, these multi-helix tiles carry multiple, programmable sticky-ends for encoding neighbor relations to dictate tile-to-tile interactions used in specific assembly of patterned superstructures. DNA tiles are formed by heating an equimolar solution of linear oligonucleotides above 90° C to melt out base-paired structures, then slowly cooling the solution to allow specific annealing to form the desired structure. Tiles are stabilized in solution by the presence of magnesium counter ions (Mg^{++}) which allow close helix packing by shielding the negative charges on the DNA backbones from one another. Design of DNA tiles and superstructures requires two separate phases: first, geometric design and second, chemical or sequence design. The geometric design phase involves modeling and examination of strand topology (paths of the oligonucleotides through the tiles), spacing of crossover points to ensure proper orientation of neighboring helical domains (for example, to ensure flatness of two-dimensional (2D) lattices), lengths of sticky-ends, and overall internal compatibility of components with each other and the superstructure design. Once the geometric constraints of the target structure are established, specific base sequences can be designed which guarantee formation of the desired structure.

Design of Base Sequences for DNA Nanoconstruction

To properly design base sequences of DNA for nanoassemblies, one must consider positive as well as negative design constraints: a sequence must not

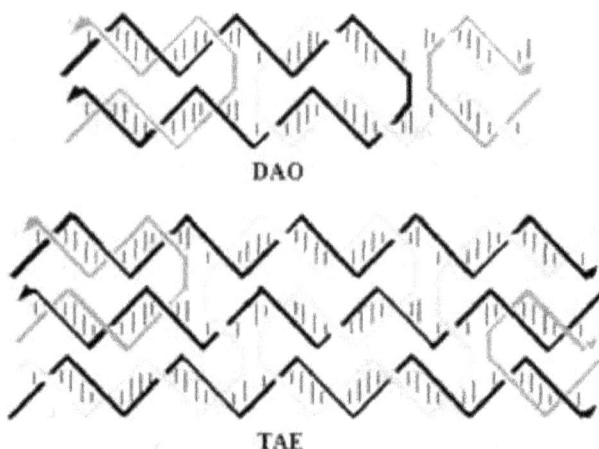

Figure 3. Example DX and TX tiles drawn as an idealized projection of 3D helices onto the plane of the page with helix axes lying horizontal on the page. Strands are shaded for ease of tracing individual oligonucleotides through the complexes. Each straight strand segment represents a half-turn around the helix. Vertical segments of strands indicate strand exchange (junction) sites where strands cross over from one helix to another. Note that two strands are exchanged at each crossover point. Arrowheads indicate 3' ends of strands. Thin vertical hashes indicate base-pairing between strands. Unpaired segments on 5' ends represent sticky-ends. The top complex is a DAO double-crossover, so called because of its Double (two) ds-helices, Anti-parallel strand exchange points, and Odd number of helical half-turns between junctions. The bottom complex is a TAE (Triple, Anti-parallel, Even number of half-turns between crossovers). Anti-parallel crossovers cause strands to reverse their direction of propagation through the complex upon exchanging helices. For example, the lightest gray strand in the DAO begins in the right-hand side of the top helix; it propagates left until it crosses over to the bottom helix, then it continues back to the right until it reaches the right-hand end of the tile. The effect of spacing between crossover points can be seen by comparing the strand trace of the DAO with that of the TAE. The TAE contains three strands (black) which span the entire width of the tile; they are the non-exchanging strands at each of the crossover points. With an odd number of half-turns between crossovers (see DAO), no strands span the width of the tile. Many other strand topologies are possible; these shown and several others have been experimentally tested. Note that the figure also shows how the minor groove of one helix is designed to pack into the major groove of neighboring helices.

only match its desired hybridization site, but it must also hold no significant complementarity to any other DNA segment, thus avoiding formation of undesired alternative structures. Many approaches and strategies for sequence design have been pursued (see for example, [Seeman, 1990; Baum, 1996; Deaton *et al.*, 1996; Marathe *et al.*, 2000; Reif *et al.*, 2001]). Primary among design constraints is Hamming distance: no sequence can be included which contains more than some threshold number of exact matches with any other sequence or the complement of any sequence already contained in the set. Thresholds are chosen based on the lengths of sequences required and known limitations from hybridization experiments. An example constraint might require at least three mismatches between every pair of subsequences of length eight. For longer strands, a sliding window is used to tabulate all subsequences of a given sequence. Such search and design problems require the use of electronic computers to keep track of the huge number of possibilities; therefore, custom software has been developed by several research groups to find good solutions to combinatorial optimization of sequence design. Besides Hamming distance, other design criteria include exclusion of certain undesired subsequences for example, palindromes which may form undesired hairpins, long stretches of G and C which, due to stronger base stacking interactions may distort the structure away from standard B-form double helix. Often, homogenization of base composition within and between strands is desirable in order to increase the likelihood of isothermal annealing. If individual regions of the structure have similar base composition they will have similar melting temperatures and formation of all parts of a tile will occur nearly simultaneously during the cooling process. Careful sequence design is critical for successful assembly of complex objects from synthetic DNA oligonucleotides since base-pair formation is the driving force of the self-organization process.

2.3 Experiments and Applications

DNA-based Computation

The first experimental proof of the feasibility of DNA-based computing came from Adleman, when he used DNA to encode and solve a simple instance of a hard combinatorial search problem [Adleman, 1994]. He demonstrated the use of artificial DNA to generate all possible solutions to a

Hamiltonian path problem (given a set of nodes connected by a set of one-way edges, answer the question of whether or not there exists a path which goes through each node once and only once). For large graphs, the problem can be very difficult for an electronic computer to solve since there are an astronomical number of possible paths and there is no known algorithm (other than complete enumeration) for finding the correct answer. Adleman's approach was to assign a 20-base DNA sequence to each node in an example graph, then to synthesize edge strands containing the complement to the 3' half of a starting node fused with the complement to the 5' half of the ending node for each valid edge in the graph. The sets of oligonucleotides encoding nodes and edges were annealed and ligated, thereby generating long DNA strands representing all possible paths through the graph. Non-Hamiltonian paths were then discarded from the DNA pool, first by size separation of the path DNA (discard strands greater than or less than the length of a Hamiltonian path, which is equal to the product of the number of nodes times the length of the node sequence), and second by a series of sequence-based separation steps involving DNA probes complementary to each node sequence (discard path sequences if they failed to contain any one of the required nodes). By this experimental protocol, Adleman was able to recover DNA strands encoding the Hamiltonian path through the example graph.

The primary contributions of Adleman's seminal paper were the revolutionary concepts that synthetic DNA could be made to carry information in non-biological ways and that the inherent massive parallelism of molecular biology operations could be harnessed to solve computationally hard problems. His experiment showed that DNA could be used as an integral part of a functioning computer. Some limits have been noted on the size of combinatorial search problems which can be implemented in DNA because of the exponential growth of search spaces and the volume constraints on wet computing techniques [Reif, 1998]. In addition to volume constraints, Adleman's original algorithm involved rather inefficient and tedious laboratory steps, the total number of which increased at least linearly with problem size. These concerns have been sidestepped by more recent theoretical and experimental advances including the development of computation by self-assembly.

Algorithmic Self-Assembly

Another fundamental insight which has shaped understanding of DNA-based computing and nanoengineering was made by Winfree when he realized that DNA annealing by itself and, specifically, annealings between DNA complexes being developed by Seeman were capable of carrying out computation [Winfree, 1998; Winfree *et al.*, 1998b]. This line of reasoning, developed theoretically and experimentally by Winfree in collaboration with Seeman and others, follows a theoretical model of computing known as Wang tiling [Wang, 1961]. In the Wang tiling model, unit square tiles are labeled with symbols on each edge such that tiles are allowed to associate only if their edge symbols match. Tiling models have been designed which successfully simulate single-tape Turing Machines and are therefore capable of universal computation [Berger, 1966; Robinson, 1971; Wang, 1975]. The recognition that DNA tiles, exemplified by DX and TX complexes (see Figure 3), could represent Wang tiles in a physical system, where edge symbols are incarnated as sticky-ends, led to proofs that DNA tilings are capable of universal computation.

Computation by self-assembly of DNA tiles is a significant advance over earlier DNA-based computing schemes because self-assembly involves only a single-step in which the computation occurs during the annealing of carefully designed oligonucleotides. Contrast this with Adleman's experiment in which the annealing step generated all possible solutions and where a long series of laboratory steps was required to winnow the set by discarding incorrect answers. Self-assembly without errors will theoretically only allow formation of valid solutions during the annealing step, thereby eliminating the laborious phase involving a large number of laboratory steps. The first report of a successful computation by DNA self-assembly demonstrated example XOR calculations [Mao *et al.*, 2000]. XOR, an addition operation without the carry-bit, was performed using tiles carrying binary values (1 or 0) designed to specifically assemble an input layer which then acted as a foundation upon which output tiles assembled based on the values encoded on the input tiles. The prototypes also demonstrated the use of readout from a reporter strand which was formed by ligation of strands carrying single bit-values from each tile in the superstructure. The scheme is currently being extended to harness the massively parallel nature of the annealing reaction by allowing random assembly of the input layers, followed by specific assembly of the output layers in order to simultaneously compute the entire lookup table for pairwise XOR (and eventually addition) up to some modest input length (perhaps 20 bits) (details described in [LaBean *et al.*, 2000a]).

Patterned DNA Nanostructures

Programmed self-assembly of DNA objects promises further advances not only in biomolecular computation but also in nanofabrication as a means of creating complex, patterned structures for use as templates or scaffolds for imposing desired structures on other materials. Simple, periodic patterns have been successfully implemented and observed on superstructures formed from a variety of different DNA tiles including DX tiles [Winfree *et al.*, 1998a], TX tiles [LaBean *et al.*, 2000b], triangular tiles [Yang *et al.*, 1998], and rhombus-like tiles [Mao *et al.*, 1999]. Figure 4 shows 2D lattice constructed from two types of TX tiles, A and B*, where the B* tiles display two extra dsDNA stem-loops (hairpins) protruding out of the tile plane, one each on the top and bottom faces of the tile. Sticky-ends on the four corners of each tile program neighbor relations such that A tiles only bind to B* tiles and vice versa resulting in the observed stripe pattern. Large lattice superstructures formed by such systems have been observed (at least 10 microns by 3 or 4 microns and containing hundreds of thousands of tiles). Larger tiles sets with more complicated association rules are currently being developed for the assembly of aperiodic patterns which will be used in the fabrication of patterned objects useful for nanotechnology applications (examples are given in Figure 5). 2D tile arrays can be thought of as molecular fabric or tapestry which contain a large number of addressable pixels. Individual tiles can carry one or more pixels depending upon the placement of observable features or binding sites. Overall connectivity can be programmed either with unique sticky-ends defined for each tile in the array or by assembly of crossover junctions which specifically stitch together distant segments of a single long scaffold strand as shown in Figure 6.

Computer simulations and theoretical analysis of self-assembly processes have pointed to some potential difficulties including the possibility of assembly errors leading to trapping of incorrectly formed structures [Reif, 1998; Winfree, 1998; Rothemund, 2000]. An experimentally observed error rate of 2-5%, encompassing annealing and ligation errors, was noted in the XOR computational complex [Mao *et al.*, 2000]. Several approaches exist to address such issues including more complicated annealing schedules, variable length sticky-ends for non-isothermal tile associations, and stepwise assembly controlled by time-stepped addition of critical oligonucleotide components. Readout methods which sample an ensemble of reporter strands as well as error-tolerant designs for the overall system are also being developed.

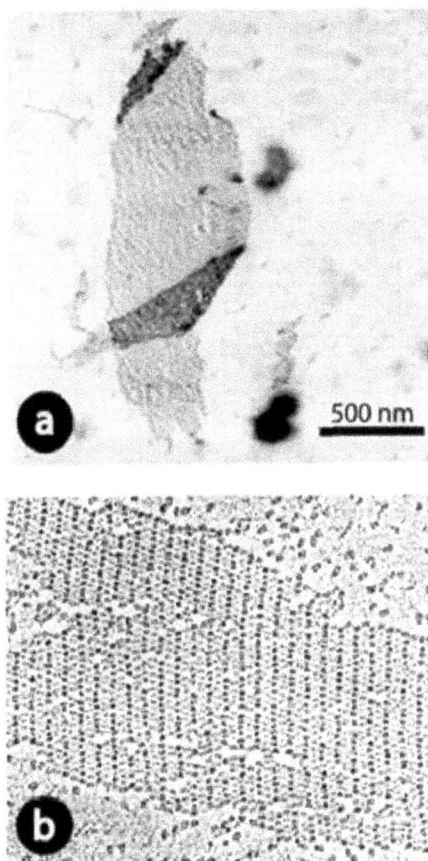

Figure 4. TX tile lattices formed by annealing eight strands and visualized by atomic force microscopy (AFM -- panel a) and transmission electron microscopy (TEM -- panel b). Lattice displaying periodic patterns (stripes in this case) was designed using two types of TX tile, A and B*. The B* tiles contained an extra hairpin of DNA projected out of the lattice plane on each side of the tile. A tiles bind only to B* tiles and vice versa by virtue of properly coded sticky-ends at the four corners of each tile. The hairpins impart distinct features which can be microscopically observed. The TEM sample (panel b) was prepared by platinum rotary shadowing resulting in the B* tiles' extra hairpin causing them to take on a darker color than the A tiles.

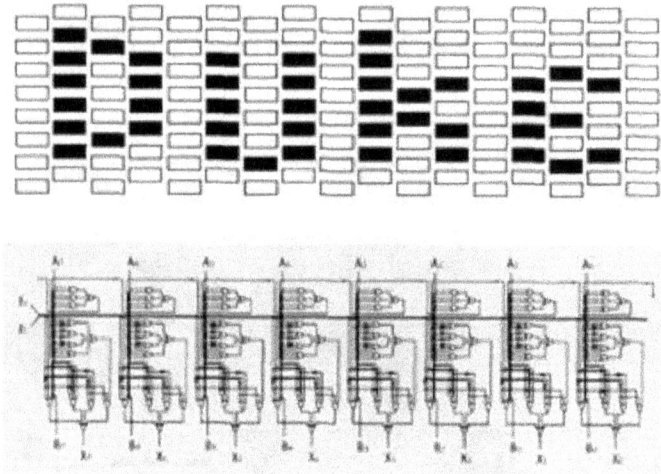

Figure 5. Examples of simple and complex aperiodic structures as possible fabrication targets for DNA-based self-assembly. A relatively simple aperiodic structure such as writing a word in addressable pixels on a DNA tile array (top) would help improve methods for eventual assembly of very complex structures such as entire circuit layouts (bottom).

Patterned Immobilization of Other Materials on DNA Arrays

Implicit in the preceding discussion of DNA self-assemblies as templates for specific patterning of other materials is the need for attachment chemistries capable of immobilizing these materials onto DNA arrays.

Materials of interest might include metal nanoparticles, peptides, proteins, other nucleic acids, and carbon nanotubes among others. A variety of strategies and chemistries are being developed including thiols (-SH), free amine groups, biotin-avidin association, and annealing of pre-labeled complementary DNA. Oligonucleotides, chemically labeled with a thiol group on either the 5' or 3' end readily bind to gold and have already been used via simple complementary DNA annealing to impart 3D ordering on gold nanospheres [Alivisatos et al., 1996; Mirkin et al., 1996; Mucic et al., 1998] and gold nanorods [Mbindyo et al., 2001]. In those studies, gold was labeled with multiple copies of a single DNA sequence, then linear dsDNA

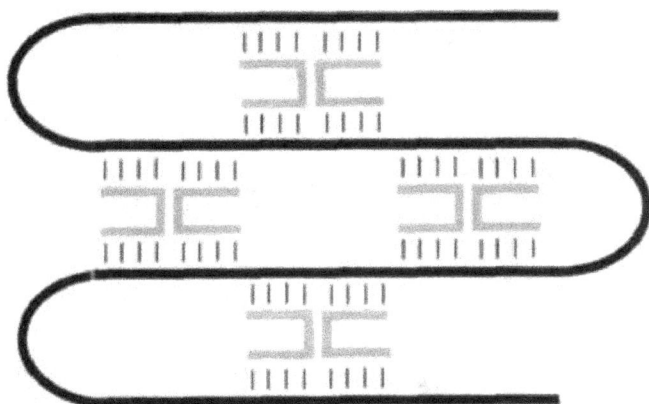

Figure 6. Schematic of a grid structure formed by annealing specific short oligonucleotides (gray) onto a preexisting long ssDNA (black). 2D arrays might be assembled not only from short synthetic strands but also making use of the larger-scale connectivity information available in long strands of ssDNA. Pixels on such a lattice would be individually addressable by virtue of their specific ordering along the large scaffold strand. Assembly of multi-tile superstructures around input scaffold strands of moderate size has been demonstrated [LaBean *et al.*, 1999]. The possibility of using very long ssDNA from biological sources is currently being investigated.

was formed between complementary strands attached to adjacent gold particles. More specific chemistries are available including nanogold reagents which make use of 1.4 nm diameter gold clusters, each functionalized with a single chemical moiety for specific reaction with a thiol or a free amino group (Nanoprobes, Inc., Yaphank, NY). These reagents have been used to target the binding of single gold nanoparticles to specific locations on DNA nanoassemblies. Figure 7 shows preliminary results of targeted binding of nanogold to a filamentous DNA tile superstructure, followed by deposition of silver onto bound gold for the fabrication of nanometer scale (~50 nm diameter) metallic wires. A similar technique has been reported for construction of a conducting silver wire on a length of ssDNA [Braun *et al.*, 1998]. Ongoing studies focus on formation of smaller (~10 nm diameter) metal wires laid out in specific patterns on 2D tile lattices.

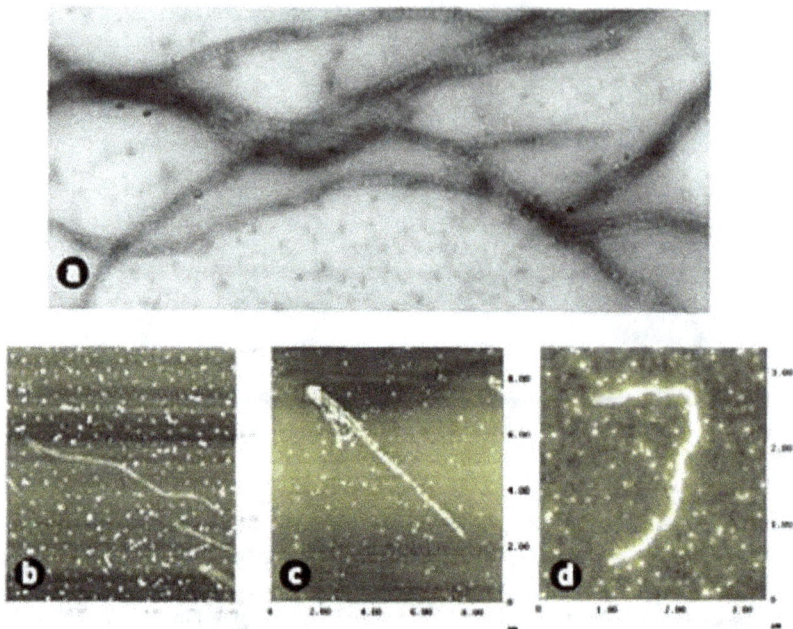

Figure 7. Targeted metallization of a complex DNA superstructure. a). Filaments of DNA lattice constructed from AB* tiles as in Figure 4 but with the addition of two thiol (-SH) groups to the dsDNA stem protruding from one side of the B* tile and one amino group on the end of the dsDNA stem protruding from the other side of the tile. It appears that the thiol sulfurs associate with one another causing the lattice to curve and form tubes of quite uniform diameter. Experiments are ongoing to further clarify details of the structure. b). Same as a) with addition of 1.4 nm nanogold targeted to the amino groups on the protruding DNA stem. c). Same as b) with addition of 2 minute development with a silver enhancement procedure which deposits silver upon existing bound gold particles. d). Same as b) but with 5 minute silver enhancement. Progressive build-up of metal atoms is observed, with perhaps a few more minutes of silver binding required to form a complete, conductive wire. Note that these DNA filaments still have sticky-ends available at both ends which can be used for orienting the entire filament prior to metal binding.

The long-term goal of these metalization studies is the self-assembly of electronic components and circuits at length scales below those available by lithography techniques.

A novel approach to targeted binding which has yet to be experimentally tested is the display of "aptamer" domains, which have been artificially evolved for specific binding of antibodies (immunoglobulin proteins) to DNA or RNA [Tsai *et al.*, 1992]. Techniques have been developed for *in vitro* selection of specific nucleic acid/antibody pairs. The antibody can be utilized as an adapter molecule, binding not only to its DNA epitope displayed on a 2D lattice but also to another protein of interest (this scheme will be further developed below). The well-known association between biotin and avidin has also been shown to be useful for targeted binding of the streptavidin protein to DNA lattice carrying an oligonucleotide labeled with the small biotin molecule [Winfree *et al.*, 1998a]. The development of these and other attachment strategies has just begun. Many advances and new insights can be expected.

2.4 Summary and Future Directions

The field of self-assembling DNA nanofabrication has already yielded successes on several fronts including binary computation, periodic tilings in two dimensions, and targeted immobilization of metallic nanoparticles. DNA has been shown to be well-suited for programmed construction of micron-scale objects with nanometer-scale feature resolution. Eventually, DNA-based self-assemblies may serve a critical role in the pattern formation step of electronic circuit fabrication, outperforming lithography by creating thousands or millions of copies of a desired structure simultaneously and at length-scales unavailable with current production techniques. Known hurdles which must be overcome include reduction of the error rate of strand hybridization, positioning of DNA objects relative to macroscopic contacts, and successful construction of complex, aperiodic patterns by algorithmic assembly.

Alternative chemistries should also be explored for backbone and bases. For example, inosine, which is able to pair with any of the four standard bases, might be useful for promiscuous annealing in some applications. Also, artificial bases with specific pairing partners could be incorporated in order to increase the information density of the polymers [Tae *et al.*, 2001]. More stable backbone variants might be investigated such as PNA, which contains the normal DNA bases linked via peptide backbone chemistry and readily forms DNA-like double helix (see for example [Hanvey *et al.*, 1992]). DNA has been exploited primarily because techniques for very specific manipulations of its base sequence and backbone connectivity have been perfected for use in recombinant molecular genetics for biotechnology applications. However, other polymers with programmable interactions might be more suitable in the long-run for some nanofabrication applications.

Possible Future Applications

Some possible fields of application for future DNA nanotechnologies might include electronic circuit lay-out, organization of materials for batteries or flat panel displays, macromolecular patterned catalysts for chemical assembly lines, combinatorial chemistry, sensorless sorting of nanometer-scale objects, DNA sequence comparison, and perhaps gene therapy.

• Electronics and Chemistry. DNA self-assemblies may find uses not only in templating nanometer scale electronic circuits alluded to in preceding sections but also in preparation of patterned catalyst arrays. For example, nanoparticulate metals used to catalyze the formation of single-walled carbon nanotubes have previously been used when randomly distributed in aerogels [Su *et al.*, 2000]. If attachment chemistries can be adapted for the binding of such nanoparticles to DNA tile lattices, then coordinated synthesis of ordered arrays of carbon nanotubes might be possible. Such ordered nanotube arrays might be useful in advanced electrical storage batteries, flat panel displays with ultra-fine pixel density, or very strong, multi-tube fibers and cables. This approach is especially attractive because current synthesis methods generally yield tangled masses of nanotubes which have been difficult to sort and organize. Other target catalysts include protein enzymes or surface catalysts which, when ordered in series, could act as macromolecular chemical assembly-lines. Patterned stripes of catalysts could act sequentially to carry out a sequence of specific reactions or even repeated cycles of reactions on a stream of substrate flowing past.

• Combinatorial Chemistry. Brenner and Lerner proposed the use of DNA for tagging chemical compounds with specific labels for use in combinatorial chemistry [Brenner and Lerner, 1992]. They suggested that DNA labels could be decoded to reveal the identity of active molecules drawn by a screening assay from a vast pool of candidate chemicals. It is possible that DNA tile structures could be used further to hold chemical reactants close together in space, thereby facilitating their reaction. The product of the reaction would remain bound to the tile, decoding of each strand of the tile would reveal the identity of each reactant used in the formation of active compounds. Encoding labels for reactants rather than final compounds would decrease the number of specific labels required.

• Sensorless Sorting. DNA tile lattices specifically decorated with protein rotary motors or environmentally responsive peptides might prove useful for sensorless sorting of poorly soluble nano-scale objects such as "buckyballs" or fragmented carbon nanotubes. Sensorless sorting involves an array of effectors capable of repetitive motion which act to organize objects into specific orientations and move them along a path comparable to a conveyor belt. Carbon nanotubes might be an interesting target object for sorting because they are poorly soluble in aqueous solution and they are difficult to purify and sort yet they are objects of intense study due to their unique structural and electronic properties. A possible scenario might involve a DNA array acting to organize a set of protein rotary motors which then provide a sweeping motion to coax nanotubes into alignment and feed them down a channel. Such an elaborate system could prove useful for simultaneously orienting large numbers of carbon nanotube into position for use as wires in a circuit, for example.

• General Nanofabrication. Self-assembling DNA-based structures also hold great potential in "seeding" for the autonomous growth of complex structures by bottom-up nano-fabrication. A molecular machine built of and fueled by DNA has been demonstrated experimentally [Yurke *et al.*, 2000]. The technique introduces the possibility of setting up a cascade of annealing reactions which, once begun, run sequentially without further intervention, and result in formation of a complex structure inaccessible by simple annealing procedures.

• Gene Sequence Comparisons. DNA is also the perfect molecule for comparison of a set of related DNA sequences. If a family of genes (e.g. analogous genes from different organisms) are annealed together with synthetic strands designed to bridge between related sequences, then the existence or the morphology of the resulting superstructure might convey information about the extent of sequence similarity in the gene set.

• Gene Therapy. It is difficult to imagine any better material for the construction of a therapeutic agent targeted toward DNA than DNA itself. Target sequence specificity is readily programmable, complex structures which bring together fairly distant regions of a long strand may be possible, stability at physiologic-like temperature and solution conditions, and the ability to organize non-DNA materials may contribute to the usefulness of DNA tiles as therapeutics. As is the case with conventional gene therapy agents, delivery may be the key limiting factor. Experiments are planned which will test the encoding of complete DNA tiles on a single cloning vector. This will not only increase the yield and decrease the cost of tiles, but it may mitigate the problem of delivery of multiple strands to a target location. A self-assembling DNA tile structure for gene therapy could make use of the fact that participation in crossover complexes increases resistance to nuclease enzymes over that of standard dsDNA. A properly delivered complex which specifically hybridizes with a target site on cellular DNA or mRNA may act to sequester the bound nucleic acids and turn off an undesired cellular response. Alternatively, if distant regions of the cellular nucleic acid were held close together within a DNA crossover complex it might be possible to activate a cellular repair mechanism and cause the excision of some portion of a faulty gene or perhaps the delivery of a corrected copy. It also might be possible to design DNA assemblies which act as diagnostics to probe for multiple mutations or multiple, specific alleles simultaneously.

Acknowledgments

Funding for some of the research described here was provided by DARPA and NSF in grants to John Reif who has been invaluable to the progress of this research. Thanks also go to Wolfgang Frey who performed the AFM shown in Figure 4, David Anderson, who performed the rotary shadowing for the sample shown in Figure 4, and Dage Liu who did the experiments summarized in Figure 7. Thanks also to Hao Yan and Lizbeth Videau for critical reading of the manuscript.

References

Adleman, L.M. (1994) "Molecular computation of solutions to combinatorial problems." *Science* **266**, 1021-1023.

Alivisatos, A.P., Johnsson, K.P., Peng, X., Wilson, T.E., Loweth, C.J., Bruchez, M.P Jr. and Schultz, P.G. (1996) "Organization of 'nanocrystal molecules' using DNA." *Nature* **382**, 609-611.

Baum, E. B. (1996) "DNA sequences useful for computation." In L.F. Landwebe and E.B. Baum, editors. *DNA Based Computers II: DIMACS Workshop* (Princeton University, June 1996) American Mathematical Society.

Braun, E., Eichen, Y., Sivan, U. and Ben-Yoseph, G. (1998) "DNA-templated assembly and electrode attachment of a conducting silver wire." *Nature* **391**, 775-778.

Brenner, S. and Lerner, R.A. (1992) "Encoded combinatorial chemistry." *Proc. Nat. Acad. Sci. USA* **89**, 5381-5383.

Chen, J-H., Kallenbach, N.R. and Seeman, N.C. (1989) "A specific quadrilateral synthesized from DNA branched junctions." *J. Am. Chem. Soc.* **111**, 6402-6407.

Chen, J-H. and Seeman, N.C. (1991) "The synthesis from DNA of a molecule with the connectivity of a cube." *Nature* **350**, 631-633.

Deaton, R., Murphy, R.C., Garzon, M., Franceschetti, D.R. and Stevens, S.E. Jr. (1996) "Good encodings for DNA-based solutions to combinatorial problems." In L.F. Landweber and E.B. Baum, editors. *DNA Based Computers II: DIMACS Workshop*, (Princeton University, June 1996) American Mathematical Society.

Du, S.M. and Seeman, N.C. (1994) "The construction of a trefoil knot from a DNA branched junction motif." *Biopolymers* **34**, 31-37.

Fu, T.-J. and Seeman, N.C. (1993) "DNA double-crossover molecules." *Biochemistry* **32**, 3211-3220.

Hanvey, J.C., Peffer, N.J., Bisi, J.E., Thomson, S.A., Cadilla, R., Josey, J.A., Ricca, D.J., Hassman, C.F., Bonham, M.A. and Au, K.G. (1992) "Antisense and antigene properties of peptide nucleic acids." *Science* **258**, 1481-1485.

LaBean, T.H., Winfree, E. and Reif, J.H. (2000a) "Experimental progress in computation by self-assembly of DNA tilings." In E. Winfree and D.K. Gifford, editors. *DNA Computers V: DIMACS Workshop June 14-15, 1999*, volume 54 of *DIMACS: Series in Discrete Mathematics and Theoretical Computer Science*, American Mathematical Society, 2000.

LaBean, T. H., Yan, H., Kopatsch, J., Liu, F., Winfree, E., Reif, J.H. and Seeman, N.C. (2000b) "The construction, analysis, ligation and self-assembly of DNA triple crossover complexes." *J. Am. Chem. Soc.* **122**, 1848-1860.

Liu, F., Sha, R. and Seeman, N.C. (1999) "Modifying the surface features of two-dimensional DNA crystals." *J. Am. Chem. Soc.* **121**, 917-922.

Mao, C., LaBean, T.H., Reif, J.H. and Seeman, N.C. (2000) "Logical computation using algorithmic self-assembly of DNA triple-crossover molecules." *Nature* **407**, 493-496.

Mao, C., Sun, W. and Seeman, N.C. (1999) "Designed two-dinesional DNA Holliday junction arrays visualized by atomic force microscopy." *J. Am. Chem. Soc.* **121**, 5437-5443.

Mao, C., Sun, W., Shen, Z. and Seeman, N.C. (1999) "A DNA nanomechanical device based on the B-Z transition." *Nature* **397**, 144-146.

Marathe, A., Condon, A.E. and Corn, R.M. (2000) "On combinatorial DNA word design." In E. Winfree and D.K. Gifford, editors. *DNA Computers V: DIMACS Workshop June 14-15, 1999*, volume 54 of *DIMACS: Series in Discrete Mathematics and Theoretical Computer Science*, American Mathematical Society, 2000, 75-90.

Mbindyo, J.K.N., Reiss, B.D., Martin, B.R., Keating, C.D., Natan, M.J. and Mallouk, T.E. (2001) *Advanced Materials* **13**, 249-254.

Mirkin, C.A., Letsinger, R.L., Mucic, R.C. and Storhoff, J.J. (1996) "A DNA-based method for rationally assembling nanoparticles into macroscopic materials." *Nature* **382**, 607-609.

Mucic, R.C., Storhoff, J.J., Mirkin, C. A. and Letsinger, R. L. (1998) "DNA-directed synthesis of binary nanoparticle network materials." *J. Am. Chem. Soc.* **120**, 12674-12675.

Peyret, N., Seneviratne, P. A., Allawi, H.T. and SantaLucia, J. Jr. (1999) "Nearest-neighbor thermodynamics and NMR of DNA sequences with A-A, C-C, G-G, and T-T mismatches." *Biochemistry* **38**, 3468.

Reif, J.H. (1998) "Paradigms for biomolecular computation." In *Proceedings of First International Conference on Unconventional Models of Computation*, Auckland, New Zealand, January 1998. Published in *Unconventional Models of Computation*, edited by C.S. Calude, J. Casti, and M.J. Dinneen, Springer Publishers, 72-93.

Reif, J.H., LaBean, T.H., Pirrung, M., Rana, V.S., Guo, B., Kingsford, C. and Wickham, G.S. (2001) "Experimental construction of very large scale DNA databases with associative search capability." In *Proceedings of The 7ᵗʰ International Meeting on DNA Based Computers,* University of South Florida, June 10-13, 2001, editors N. Jonoska and N.C. Seeman.

Rothemund, P.W.K. (2000) "Using lateral capillary forces to compute by self-assembly." *Proc. Nat. Acad. Sci.* **97**, 984-989.

Seeman, N.C. (1982) "Nucleic acid junctions and lattices." *J. Theor. Biol.* **99**, 237-247.

Seeman, N.C. (1990) "De novo design of sequences for nucleic acid structural engineering." *Journal of Biomolecular Structure and Dynamics* **8**, 573-581.

Su, M., Zheng, B. and Liu, J. (2000) "A scalable CVD method for the synthesis of single walled carbon nanotubes with high catalyst productivity." *Chem. Phys. Letts* **322**, 321-326.

Tae, E.L., Wu, Y., Xia, G., Schultz, P.G. and Romesberg, F.E. (2001) "Efforts toward expansion of the genetic alphabet: replication of DNA with three base pairs." *J. Am. Chem. Soc.* **123**.

Tsai, D.E., Kenan, D.J. and Keene, J.D. (1992) "In vitro selection of an RNA epitope immunologically cross-reactive with a peptide." *Proc. Nat. Acad. Sci., U.S.A.* **89**, 8864-8868.

Watson, J.D. and Crick, F.H.C. (1953) "Molecular structure of nucleic acids: a structure for deoxyribose nucleic acid." *Nature* **171**, 737-738.

Winfree, E. (1998) *Algorithmic self-assembly of DNA*. Ph.D. Thesis, Caltech.

Winfree, E., Liu, F., Wenzler, L.A. and Seeman, N.C. (1998b) "Design and self-assembly of two-dimensional DNA crystals." *Nature* **394**, 539-544.

Winfree, E., Yang, X. and Seeman, N.C. (1998a) "Universal computation via self-assembly of DNA: Some theory and experiments." In L.F. Landweber and E.B. Baum, editors. *DNA Based Computers II: DIMACS Workshop*, (Princeton University, June 1996) American Mathematical Society.

Yang, X., Wenzler, L.A., Qi, J., Li, X. and Seeman, N.C. (1998) "Ligation of DNA triangle containing double crossover molecules." *J. Am. Chem. Soc.* **120**, 9779-9786.

Yurke, B., Turberfield, A.J., Mills, A.P., Simmel, F.C. and Neumann, J.E. (2000) "A DNA-fuelled molecular machine made of DNA." *Nature* **406**, 605-608.

Zhang, Y. and Seeman, N.C. (1994) "The construction of a DNA truncated octahedron." *J. Am. Chem. Soc.* **116**, 1661-1669.

Author's Address

Thomas H. LaBean, Department of Computer Science, Duke University, USA. Email: thl@cs.duke.edu.

Chapter 3

Mapping Sequence to Rice FPC

Carol Soderlund, Fred Engler, James Hatfield, Steven Blundy,
Mingsheng Chen, Yeisoo Yu and Rod Wing

3.1 Introduction

In the late 1990's, there were discussions on whether to build physical maps to select clones for sequencing [Green, 1997] or to use a whole genome shotgun strategy [Weber and Myers, 1997]. A draft sequence of the human genome was published by the International Sequencing Consortium [2001] which was based on the human FPC (FingerPrinted Contig) map by the International Mapping Consortium [2001], and a draft sequence was published by Celera which was based on the whole genome shotgun strategy and included the draft sequence from the public consortium [Venter *et al.*, 2001]. The current general attitude is that the best approach is a combination of the two. Regardless as to whether a map is essential for sequencing, it provides a mechanism for tying together information gathered over the years, i.e. genetic, physical and sequence information. It provides a tremendous amount of locational and comparative information without having to sequence. Many large genomes will not be sequenced anytime soon as the cost is still prohibitive, yet the cost of mapping is acceptable. Currently, the price of sequencing a genome is about 3 cents per base, so approximately $4500 for a 150 kb clone, whereas fingerprinting a BAC clone is approximately $5. If an organism has a physical map with landmarks such as genetic markers and ESTs, sequencing can be restricted to the interesting regions. As sequences become available, they can be consolidated and organized along the map, as will be described in this chapter.

Over a decade ago, the first contigs built by restriction fragment fingerprints were published. Coulson *et al.* [1986] used the end-labelled double digest method with cosmid clones for mapping the 100 Mb *C.elegans* genome. Olson *et al.* [1986] used the complete digest method with lambda clones for mapping the 40 Mb yeast genome. Both genomes were subsequently sequenced based on the map. In both cases, the building of the map was largely interactive for the following reasons: First, there were many gaps as the clones were relatively small; i.e. lambda clones are about 15 kb and cosmid clones are about 40 kb. Second, there was a large amount of error and uncertainty in the data that makes automatic assembly difficult. Last, the problem is NP-hard and not near enough resources went into finding a computational solution. There were other attempts in the early 1990's to use this approach, but they also suffered from these problems. Obviously, this would not scale up to the 3000 Mb human genome. Hence, the method was thought to be unusable.

Alternative methods were suggested, such as sequencing the ends of large insert clones (referred to as STC for Sequence Tagged Connector, or BES for BAC End Sequence). When a new clone is sequenced, the sequence can be compared against the STCs to find the next clone to sequence [Venter *et al.*, 1996]. A whole genome shotgun strategy was suggested, where forward and reverse reads are taken from 2 kb and 10 kb clones, and the sequence contigs are ordered based on information from the orientation and distance between reads and from STC sequences [Weber and Myers, 1997; Myers *et al.*, 2000].

Meanwhile, the Sanger Centre was still building fingerprinted contigs using the double digest method [Bentley *et al.*, 2001] and the FPC (FingerPrinted Contigs) program was developed for this effort [Soderlund *et al.*, 1997; Soderlund *et al.*, 2000]. FPC has the combination of automation and interactive graphics. It tolerates varying amount of data where the better the data -- the better the map, it flags potential incorrect contigs, and it can assemble large numbers of clones. BACs were used for fingerprinting so there are fewer gaps as the length of a BAC is approximately 150 kb. Marra *et al.* [1999] undertook to fingerprint the whole *Arabidopsis* genome using the complete digest method using techniques that produced a large reduction in error and uncertainty in the data. Since then, chromosome 2 and 3 (80% of the genome) of *Drosophila* [Hoskins *et al.*, 2000] has been mapped, a whole genome human map [The International Mapping Consortium, 2001] and a whole genome rice map [Wing *et al.*, 2001] have been built; in all cases FPC was used. Mouse, zebrafish, and maize are now being mapped, along with many other genomes. In summary, the combination of longer clones, less error and uncertainty, and robust software has rejuvenated this method.

The advantages of having a map for the plant community is tremendous as many of the plant genomes have a much higher complexity than the human genome. Their genomes tend to be larger, more repetitive, and they can have multiple distant genomes within the nucleus. For example, maize has a haploid genome size of 2500 Mb and 60-80% of the maize genome is composed of highly repetitive retrotransposons [San Miguel *et al.*, 1996]. Barley is a diploid and has a genome size of 5000 Mb. Nearly 90% of the barley genome is composed of repetitive DNA and only one type of retrotransposon (BARE-1) constitute 2.8% of the barley genome [Vicient *et al.*, 1999]. Wheat is an allohexaploid with genome constitution AABBDD and has a genome size of 16000 Mb. It was formed through hybridisation of AA with a B genome diploid, and the subsequent hybridisation with a D genome diploid [Devos and Gale, 1997]. Table 1 shows a sample set of genomes, sizes, percent repetitive and polyploid. Even if there were the funds to sequence these genomes, it would be difficult with a whole genome shotgun approach exclusively, i.e. without an underlying map.

Arabidopsis has been physically mapped [Marra *et al.*, 1999] and sequenced [The Arabidopsis Genome Initiative, 2000]. At CUGI (Clemson University Genome Institute), we have built a physical map of rice [Chen *et al.*, 2002] to aid the sequencing of rice in collaboration with the International

Genome	Size(Mb)	Repetitive	Description
Arabidopsis	125	14	Diploid
Rice	380	76	Diploid
Maize	2500	83	ancient tetraploid
Barley	5000	88	Diploid
Wheat	16000	88	Hexaploid

Table 1. Attributes of a few plant genomes.

Rice Genome Sequencing Project (IRGSP). The sequence from these model genomes will be used in comparative analysis with other plant genomes that are not being sequenced. Regardless as to whether a plant genome will be totally sequenced, partially sequenced, or only have small pieces of sequence information available such as ESTs, the ability to map the sequence to the physical map is valuable. To aid this mapping, STCs are often generated for the clones in a fingerprinted map as this can provide a fairly even distribution of small pieces of sequence over the map. The STCs are used to map clone sequence [Hoskins *et al.*, 2000] and marker sequence [Yuan *et al.*, 2001] to the physical map.

Existing sequence can be used to anchor contigs, close gaps and verify contigs. Any BAC genomic sequence can be mapped back to the FPC map in one of three ways: (1) FSD (FPC Simulated Digest) will digest a sequence and convert it to migration rates such that it can be incorporated into the map as a fingerprint. (2) BSS (BLAST Some Sequence) blasts a clone sequence against the STC database and the sequence can be added as an electronic marker attached to all the clones to which it had a high hit with the STC. (3) BSS blasts a marker sequence against the STC or clone sequence database and the marker can be added as an electronic marker attached to all the clones to which it has a high score. All of these new features are being used extensively to complete our rice physical map. We display our contigs on the Web using a java program called WebFPC. A brief overview of FPC will be given, then a description of each of these features and results from our rice project.

3.2 Overview of FPC

For a detailed description of the algorithm, see [Soderlund *et al.*, 1997]. For simulation results, see [Soderlund *et al.*, 2000]. The following gives a brief overview. FPC (FingerPrinted Contigs) assembles clones into contigs using either the end-labelled double digest method [Coulson *et al.*, 1986; Gregory *et al.*, 1997] or the complete digest method [Olson *et al.*, 1986; Marra *et al.*, 1999]. Both methods produce a characteristic set of bands for each clone. To determine if two clones overlap, the number of shared bands is counted where two bands are considered 'shared' if they have the same value within a tolerance. The probability that the N shared bands is a coincidence is computed, and if this score is below a user-supplied cutoff, the clones are considered to overlap. If two clones have a coincidence score below the cutoff but do not overlap, it is a false positive (F+) overlap. If two clones have a coincidence score above the cutoff but do overlap, it is a false negative (F-) overlap. It is very important to set the cutoff to minimise the number of F+ and F- overlaps.

A FPC *complete build* bins clones into transitively overlapping sets where each clone in a set has an overlap with at least one other clone in the set and no clone has an overlap with any clone outside the set. The clones in a bin are given an appropriate ordering by building a CB (consensus band) map and the CB map is instantiated as a contig. Hence, a complete build guarantees that each contig is a transitively overlapping set of clones based on a given cutoff. The length of a clone in a contig is equal to the number of its bands, and the overlap between the coordinates of the two clones is approximately the number of shared bands. If clone C_A has exactly or approximately the same bands as clone C_B, C_A can be *buried* in C_B and C_B will be called the *parent*. Clones that do not have an overlap with any other clone are not placed in a contig and are called *singletons*. Markers can be attached to a clone and are displayed in the contig with the clone. A clone can only be in one contig, but a marker can be attached to clones in multiple contigs (e.g. duplicated locus). An externally ordered subset of the markers can be input into FPC as the *framework*. Contigs containing these markers can be listed by framework order in the project window. Briefly, the following are some of the most salient features of FPC:

CpM (Cutoff plus Marker): FPC provides the option of defining a set of rules on what constitutes a valid overlap, which are entered into the CpM table. For example, the table can be set so that two clones will be considered to overlap if they (i) have less than a 1e-12 score, (ii) share at least one marker and score less than 1e-10, (iii) share at least two markers and score less than 1e-09, or (iv) share at least three markers and score less than 1e-08.

IBC (Incremental Build Contigs): The IBC routine automatically adds new clones to contigs and merges contigs based on the cutoff and CpM table, and then the clones in each modified contig are re-ordered by executing the CB algorithm. The IBC provides a summary of the modifications performed on each contig in the project window.

Q clones: If there is a severe problem aligning the bands of a clone to the CB map, it is marked as a Q (questionable) clone. If there are many Q clones in the contig, the simulations show that this generally indicates at least one F+ overlap and the ordering will almost certainly be wrong. Interactive tools are available to fix these contigs.

Merge: Due to the uneven coverage of restriction fragments and the random picking of clones, there is an uneven coverage of the clones so that they assemble into many contigs. Contigs can often be merged by querying the end clones of a contig. Interactive tools are available to detect and merge contigs.

The simulations verify that the better the data -- the better the map. With a set of simulated clones from 110 Mb of human sequence, a simulated digest using *EcoRI* was performed. The largest contig assembled has 4783 clones with two out-of-order pairs, that is, when clone A should start before clone B but clone B starts before clone A, though they do correctly overlap. As error is added, the number of out-of-order pairs increases.

3.3 Mapping Sequence to FPC contigs

The following three sections describe new software developments to aid mapping and display of sequence on a FPC map.

FSD (FPC Simulated Digest)

FSD is a supplemental program (see Figure 1) to FPC that performs a complete digest *in silico* on a sequence that produces the sizes of the fragments. The sizes are converted into migration rates so that they can be assembled into the FPC map. Note that FPC can use either sizes or migration rates for each clone fingerprint. Generally, migration rates are used for FPC maps as they represent the bands on the gel image. The bands are assigned migration rates and then converted into sizes by Image (see http://www.sanger.ac.uk/Software/Image). The Human Mapping Consortium digested sequence *in silico* into fragment sizes, but did not further convert them into rates; hence, they maintained two FPC files, one in rates and one in sizes [The Human Mapping Consortium, 2001]. We have taken the extra step to convert the sizes into migration rates so that we only need to maintain one FPC file.

There were two main reasons for developing FSD. First, we wanted a way to verify both the fingerprints and the final sequence assembly. By simulating a complete digest on the final sequence, we should get a set of bands that closely match the fingerprint produced in the laboratory. This simulated fingerprint should automatically be positioned very close to the lab

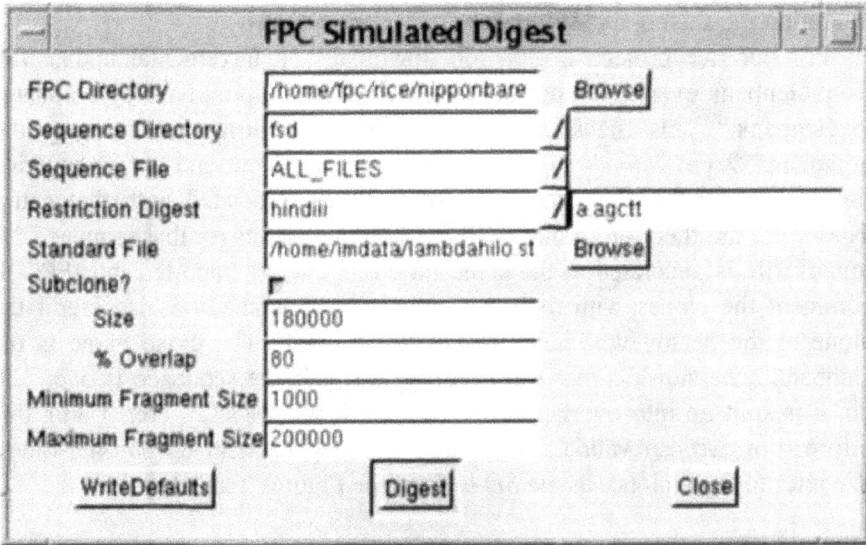

Figure 1. FSD window. FSD is a stand alone tool that takes as input one or more sequences and outputs the band and size files in a FPC format.

fingerprint. If the simulated fingerprint is very different from the lab fingerprint, this could possibly indicate misnamed clones or an incorrect sequence assembly. The second main motivation is the large amount of data publicly available from Genbank, where a percentage of the sequenced clones are not from our FPC map. With this sequence data, many new fingerprints can be generated. By adding *in silico* fingerprints from sequences generated at other labs, we would confirm our contig assembly, join additional contigs in FPC, anchor more contigs, and provide an integrated map of sequence from many sources.

FSD will take as input one or more sequences, producing bands and sizes files. The sizes file is a list of resulting fragment sizes when a sequence file is cut using a specified restriction enzyme. In order to convert the sizes to migration rates, the *standard file* is used. The standard file is created at the beginning of the fingerprinting project. When a gel is run, the set of standard markers (i.e. fragments) are also run; these markers have known rates and sizes so that the rates of the new clones can be normalized by Image. FSD fits a cubic spline curve to the standard values. It then converts the sizes to migration rates using this spline curve.

For our rice project, a cron job downloads an incremental update file from Genbank every evening that contains all of the previous day's updates to Genbank. This file is scanned for Genbank entries pertaining to the organism '*Oryza sativa*'. These entries are parsed out and put in separate file, named by the Genbank accession number associated with that entry. These files are then run through FSD to generate clones for that sequence. A remark file is generated at the same time that can be imported into FPC to comment the clones with their associated chromosome and also credit the clone to the person who submitted it to Genbank. The clone name is the Genbank accession number followed by "sd1"; if the sequence is over 180 kb, it is split up into overlapping sequences labelled "sd2", etc. Using this information we can validate clone and contig placement on chromosomes. We refer to these clones as the *SD clones* (see Figures 3 and 4).

BSS (BLAST Some Sequence)

Given that the clones in an FPC map have STCs, sequence can be mapped to the clones in the following two ways: The next clone for sequencing is selected by comparing the STCs with a new sequence, finding the one closest to the end of the clone and verifying the results by looking at the gel image [Hoskins *et al.*, 2000]. Sequence from markers has been compared to STCs to anchor contigs [Yuan *et al.*, 2001]. In both cases, much of this process is automated by BSS, saving the biologist time spent examining results, and allowing more experimentation with search parameters. BSS uses the popular BLAST software [Altschul *et al.*, 1997], which provides results in a format that the biologist is familiar with. In addition to mapping sequence and markers to the STCs, the BSS allows mapping of marker sequence to genomic sequence associated with clones in the map. The BSS mappings are summarized in Table 2.

These mappings can be run on a sequence associated with a clone in a contig or on a directory of sequences. The database sequences (STC or genomic) must be associated with clones in FPC; this association is done by having the FPC clone name be a substring of the STC or genomic sequence name. The function can be run in *contig mode*, in which only the sequence within a contig is searched, or in *batch mode*, in which all sequences in

Query	Database
Sequence	STC
Marker	STC
Marker	Sequence

Table 2. BSS mappings of Query→Database.

the database are searched. In this section, we will look at three tasks that can benefit from such procedures.

Picking a minimal tiling path (MTP)

For this task, a minimally overlapping set of clones is selected for sequencing. Note that it costs a lot of extra effort if there is a gap between two supposedly overlapping clones, or if their overlap is large causing too much redundant sequencing. The MTP clones may be picked by viewing the fingerprints of the adjacent clones. The benefits of this method are that it gives orientation information and many clones can be selected without having to wait for the sequence of a clone. The disadvantage is that a large piece at the end of a clone may not be detected by electrophoresis; by only looking at the fingerprint, what appears to be a minimally overlapping clone may actually overlap a lot. Another approach is to query a database of STCs with a sequenced clone to determine the next clone to sequence. The advantage of this method is that when a hit is found near the end of a clone, the overlap will probably be minimal. The disadvantage is that when run against all the STCs, this produces a large number of false hits that the user must filter through. These false hits occur due to the presence of repetitive sequence and there is error in the STC sequence as it is single pass sequence. A second problem is that a large number of STCs are misnamed. A third problem is that this approach does not show orientation, that is, a STC may hit near the end of the clone but whether it extends away from the clone or into the clone is often not obvious. It is therefore vital to confirm these hits by linking each hit to a clone and viewing the results on the physical map.

With BSS, this is done by selecting a Sequence→ STC mapping in contig mode, setting the query to the clone's sequence file, and setting the database to the STC library. After setting any desired BLAST parameters, a *Current Contig* search is performed to search the STCs of clone within the current contig, and/or a *Contig Ends* search is performed to search the STCs from all clones at the end of contigs. A summary of the resulting hits and their quality is provided in the BSS Results display and the exact alignment details may also be viewed. If the hits are deemed significant, they may be added to the FPC map either as electronic markers or as remarks. From these results, one can easily select the clone with minimal overlap and confirm the overlap by looking at the fingerprints.

Merging contigs

Often, fingerprints do not give enough information to identify overlapping clones. Even with 20x coverage, this problem still exists since usually 70% of the bands must be shared between two clones to rule them as overlapping. Because of these apparent gaps in the map, the physical map of a single contiguous segment often takes the form of several contigs. These contigs must be manually examined to determine which contigs should be merged. Analyzing the fingerprints close to the ends of contigs with a less stringent cutoff is generally used to determine which contigs to merge. Furthermore, sequencing a clone close to the end of contig, querying it against a STC database, and looking at hits close to the ends of contigs provides additional, more fine-grained information. For this task, BSS helps us identify potential merges. If significant STC hits occur in another contig, that contig may be merged with the current one. The setup is identical to the one used when picking a MTP. However, a *Contig Ends* search will always be performed and using the batch mode allows many sequences to query the database. Figure 2 shows an example of BSSing a directory of 5x draft sequences against the STC database.

Anchoring contigs

When sequence is associated with genetic markers, the markers may be placed on the map electronically by querying the STC or clone sequence database for matches to the marker sequence and positioning the marker where hits occur. If we wish to query the STC database for hits, we will

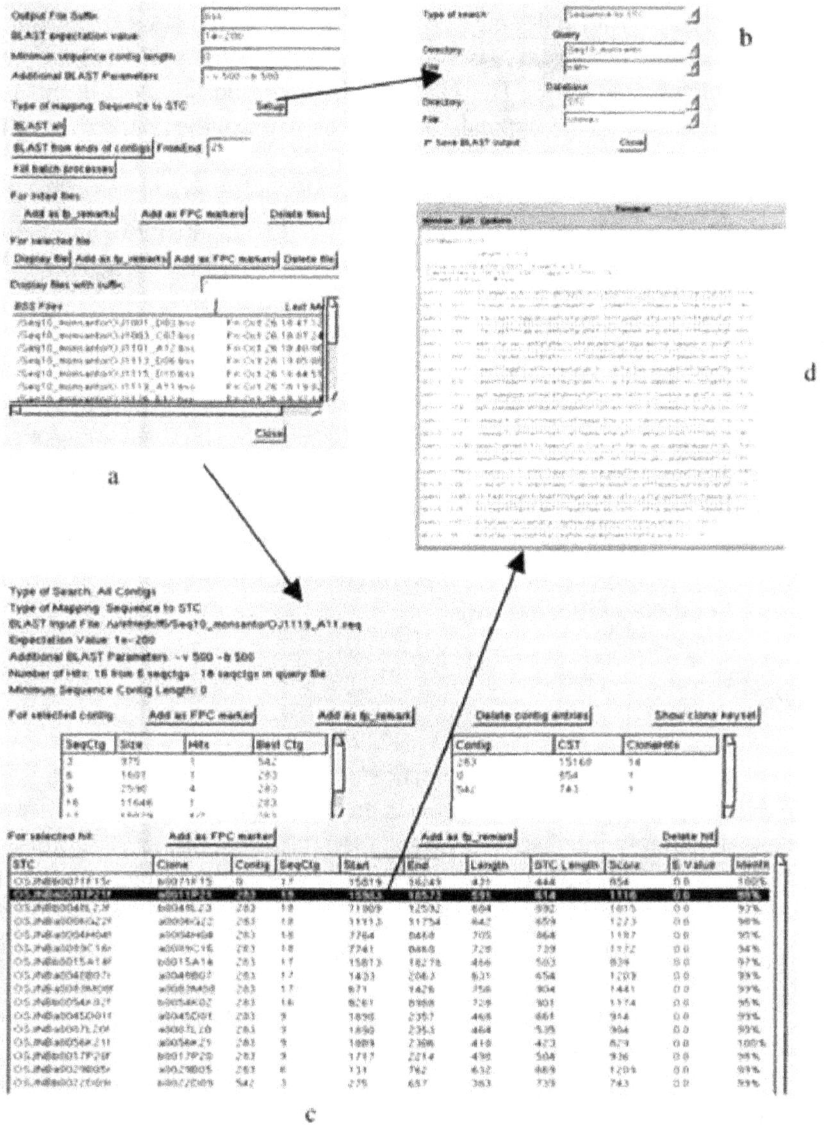

Figure 2. BSS windows. (a) The driver window for running BSS in batch mode. A list of the result files is shown at the bottom. (b) The setup window for selecting function, directories and files. (c) A selected results file is shown in the results window for a Monsanto sequence file of 5x coverage which assembled into many sequence contigs, referred to as SeqCTG. (d) The alignment of one SeqCTG to an STC.

select the Marker→STC mapping. If we wish to query a set of sequences with corresponding clones in the physical map, we will select the Marker→Sequence mapping. The batch mode would be the most useful for this option, as it would be typical to want to map all the markers to any contig.

The following scenario gives us an example of an application for the Marker→Sequence mapping. Sequence is downloaded from GenBank via the Internet, and band files are created from the sequence using FSD (previously described), which allows us to add the sequence as clones to the FPC map. Marker sequences then search these clone sequences for hits. If significant hits should occur, we can anchor contigs based on the information. Most importantly, all of these steps can be performed without any lab work.

All three problems just described demonstrate the advantages of integrating the physical mapping approach with sequence comparisons. By filtering out unwanted hits, and allowing the user to view results in light of a physical map, BSS effectively reduces the tedious work of sorting through pages of results, and opens up new opportunities for solving problems arising during map building and sequencing.

WebFPC

FPC is a very powerful and sophisticated program. However, there are some users for whom all this power is far more than they need. These researchers are simply interested in viewing the data, nothing more. For these researchers, WebFPC was created. Written as a Java applet, WebFPC provides the user with an easy way of viewing physical maps simply by clicking on a link in a web page; see Figure 3 for an example. The applet locates and downloads the data automatically. Because of the amount of data involved, a few server side scripts have been developed to separate and compress the data to speed up download time.

In order to integrate WebFPC maps with other relevant databases around the world, we have set up two mappings (see Table 3 for rice mappings). The first allows any external site to start up the Java Applet for a particular contig with a given clone or marker highlighted. The second allows any external site to send us a file of clones and/or markers for which they have Web-based information, a URL and a database name. We only need to add their file to a directory of files and run a script. Thereafter, their database will be listed on the database pull-down button for a contig, and when a clone or marker is

Rice WebFPC	Links: Genbank. Gramene	http://www.genome.clemson.edu/projects/rice/fpc
Rice Status	Links: WebFPC	http://www.genome.clemson.edu/projects/rice/ccw
Gramene	Links: WebFPC	http://www.gramene.org

Table 3. URLs for Rice FPC.

selected, it will say if there is a clone or marker in the external database, and if so, the user can request it.

Results from Our Rice FPC Map

Rice FPC has 68k clones (~20x coverage) from two BAC libraries, one cut with *Eco*RI and the other cut with *Hind*III. We have 1202 markers and 706 genetic markers from the Japanese High Density Genetic Map [Sasaki and Burr, 2000]. By hybridising genetic markers to clones, the contigs are ordered and anchored to the chromosomes. Approximately 90% of the genome is covered by anchored contigs. We also have STCs for about 80% of our BAC clones. The CCW consortium (CUGI: Clemson University Genome Institute, CSHL: Lita Annenberg Hazen Genome Sequencing Center at Cold Spring Harbor Laboratories, GSC: Washington University Genome Sequencing Center) are sequencing and annotating the short arms of chromosomes 3 and 10.

A total of 346 sequences from chromosome 1 have been submitted to Genbank by the RGP (Rice Genome Program, http://rgp.dna.affrc.go.jp) as of September 2001. These clones are not from our rice FPC but are BACs and PACs from the RGP minimal tiling path. These have been downloaded, run through the FSD program, and added to the rice FPC file automatically. The map locations of the SD fingerprints were in agreement with the chromosome anchoring and marker orders determined during physical map construction of

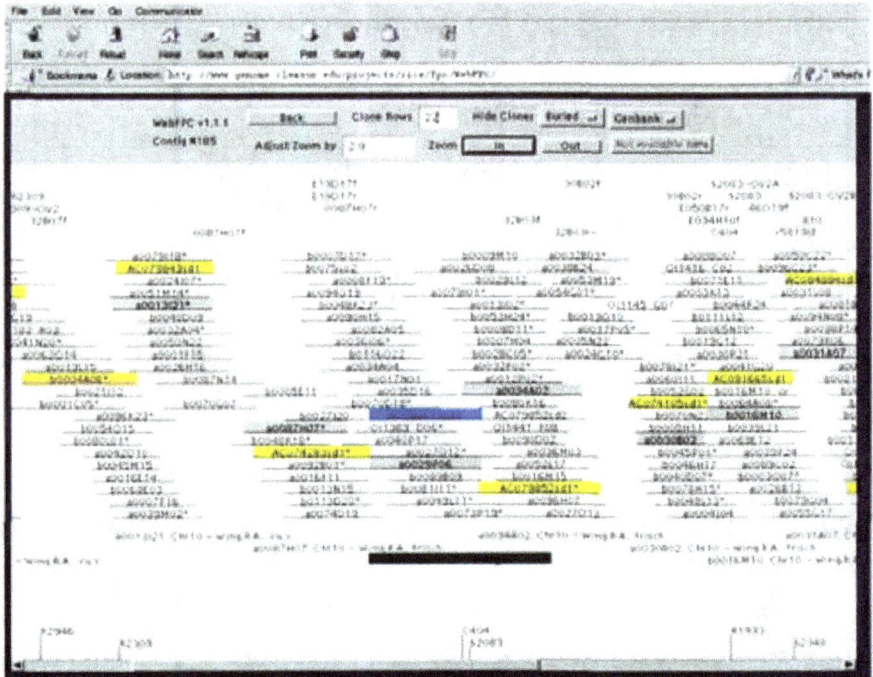

Figure 3. WebFPC with SD clones. The clones ending in 'sd' are from digesting *in silico*. They are coloured yellow to represent finished clones; the grey clones are the corresponding true fingerprints.

their contig for 305 of these fingerprints, leaving 41 clones as singletons. Of these 41 clones, 23 could be positioned correctly by lowering the cutoff. Of the remaining 18 clones, 12 were located in low coverage regions, 4 were too small to match standard size clone fingerprints, one clone was misassembled, and one clone mapped to the wrong location. The WebFPC display in Figure 4 shows the minimal tiling path of a subset of the Japanese clones. A total of 1352 rice sequence clones have been downloaded from Genbank and can be viewed from the WebFPC for rice; see Table 3 for the URL.

As mentioned, we anchor contigs based on the Japanese High Density Genetic Map. FPC takes as input a framework file of ordered markers with their locations. This function was written for the chromosome specific Sanger Centre maps. We use it for a whole genome map as follows: The

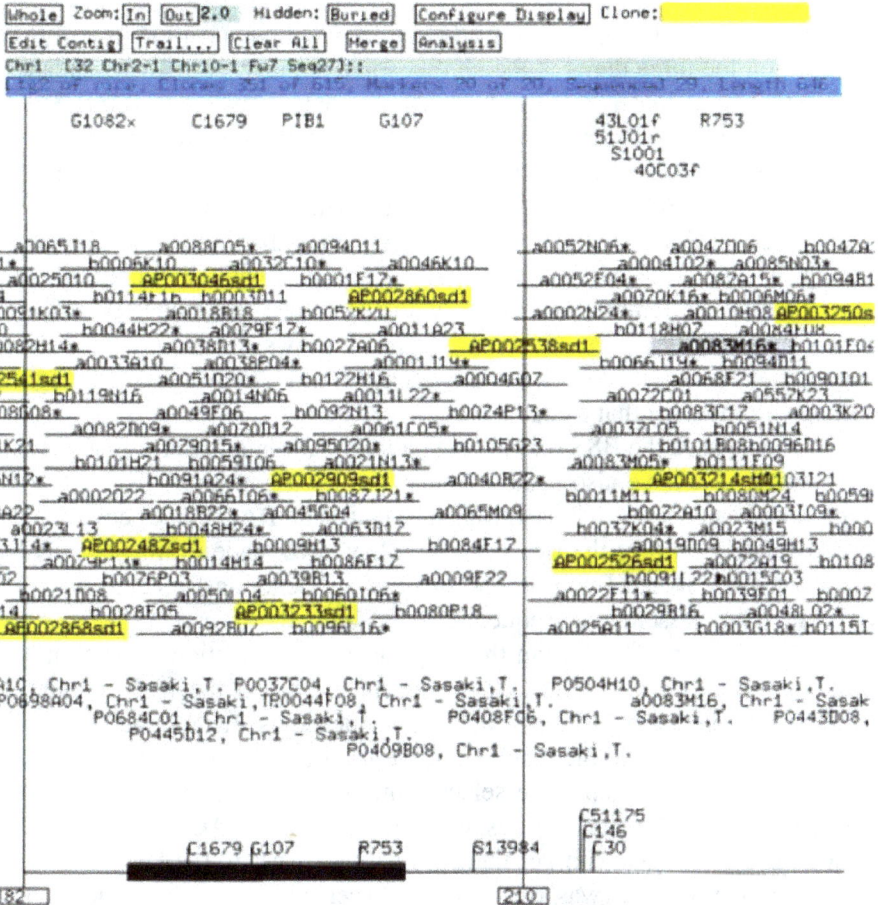

Figure 4. Anchored contig. The clones ending in 'sd' are from digesting *in silico* the Japanese clones. In the contig remark, additional information is given as to the amount of evidence for the anchoring:

  Chr1 [32 Chr2-1 Chr10-1 Fw7 Seq27] ::

indicates that this contig has 7 framework markers and 27 SD clones giving a total of 34 hits, where 32 of the hits were on chromosome 1, and the other two hits were on chromosome 2 and chromosome 10. Note that the "Update ChrN contig remark" in FPC puts the remark at the beginning of the contig remark. Anything before a "::" does not get removed unless explicitly requested. Automatic remarks are added after the "::".

location can be three digits followed by a digit and one position of accuracy. Proceeding the three digits is the chromosome number, e.g. 1001.3 indicates the marker is at location 1.3 cM on chromosome 1, whereas 10001.3 is at location 1.3 cM on chromosome 10. The SD sequence provides a second way to anchor contigs and verify existing anchored contigs. A routine has been added to FPC, located on the Project Window/Menu Window, button name "Update ChrN contig remark", which does the following: For each contig, it counts the number of anchors associated with each chromosome, and it parses the remarks associated with SD clones and counts the number associated with each chromosome. Figure 4 shows a contig with a ChrN contig remark, where there are ambiguities. WebFPC only shows the highest hitting chromosome that a contig hits.

We have used the BSS for selecting a minimal tiling path for rice genome sequencing. We have also used it to map the Monsanto draft [Barry, 2001]. Monsanto has generated 5x coverage of about 3000 BAC clones covering the approximately 50% of the rice genome. They have made available to us the sequence files. Robin Buell of The Institute of Genomic Research provided us with 303 assembled sequence files and BAC end sequences on chromosome 3 and 10. We ran the Sequence→STC function in batch mode, which mapped all the draft to our sequence. Due to the high amount of repetitive sequence, we did not add the sequence as markers as it was too much data. By viewing the result files, we used it to help select the minimal tiling clones and fill gaps. We selected minimal tiling BACs by examining the BSS output of distal contigs in each Monsanto BAC. The amount of sequence overlap (one can get the information based on the STC alignment from the BSS output) was calculated between two BACs and then a few candidates' fingerprints were compared from adjacent clones to select the best clones to be sequenced. We successfully selected more than 30 clones on chromosome 3 with the BSS function. Moreover, in some cases, we identified a sequence contig that contains STCs of two adjacent sequenced BACs. The direction of STC alignments was compared to the FPC clone order to confirm the possibility of gap or overlap. Simply, if the direction of STC alignments point towards each other, then it is an overlap and if STC alignments point in different directions, then it is a gap. By doing this, we closed three gaps (around 10kb or less) on short arm of chromosome 10.

Table 3 shows the URLs to web based Rice FPC maps. The Chromosome 10 status page has links to the WebFPC map. WebFPC has links from SD clones back to the original Genbank record and to the Gramene clone description [Ware *et al.*, 2002]. And we are working with other groups to link databases, e.g. the Gramene map, which is a comparative mapping resource for grains, links back to the Rice FPC map.

3.4 Discussion

Sequences from various sources are being generated and this sequence can be mapped to the FPC map using the FSD and BSS tools. Additionally, the FSD software helps validate clones, merge contigs and anchor contigs. The BSS software helps select a minimal tiling path, merge contigs, and we are now using it to further anchor contigs. The genome produces massive amounts of data. It takes time and energy to consolidate the data, and doing the mundane parts of this work is prone to error. Much less, the results are in many places. The mapping of sequence to a FPC map using FPC compatible functions will reduce error and make the ability to do the mapping available to many laboratories, even those with small to non-existent bioinformatics staff. The WebFPC allows everyone to view our data and link with other databases, hence, greatly supporting collaborative efforts.

The mapping and sequencing of rice is an international effort, and our software development over the last year greatly aids this international effort by consolidating and displaying data, as follows: we use the Japanese High density map to order our contigs, and then add clones from around the world through the SD (simulated digest) clones, and display the integrated rice map on WebFPC. We are now using BSS to map more of the 3267 Japanese High Density markers to our map based on the sequence of these markers; when a new marker gets added to FPC that is in the framework map, the contig gets automatically anchored. We will then proceed to map ESTs from various plants to our map, which will work exactly the same as mapping the genetic markers.

Mapping ESTs, marker sequence, and genomics sequence from other genomes will be a great aid to comparative genomics. The genomes will not need to be sequenced completely in order to get valuable cross information. And as discussed in the introduction, this is exceptionally valuable to plant genomes, as various laboratories generate regional or function specific sequences, which can be placed on a global FPC map.

Acknowledgments

CUGI was funded by Novartis to fingerprint the Rice nipponbare BAC libraries. The CCW (CUGI, CSHL, WashU) consortium of Clemson University Genome Institute, the Lita Annenberg Hazen Genome Sequencing Center at Cold Spring Harbor Laboratories, and the Washington University Genome Sequencing Center was awarded a grant from the USDA - CSREES/NSF/DOE rice genome initiative to sequence and annotate the short arms of chromosomes 3 and 10.

References

Altschul, S., Madden, T., Schaffer, A., Zhang, J., Zhang, Z., Miller, W. and Lipman, D. (1997). "Gapped BLAST and PSI-BLAST: A new generation of protein database search programs." *Nucleic Acids Res.* **25**, 3389-3402.

Barakat, A., Carels, N. and Bernardi, G. (1997). "The distribution of genes in the genomes of Gramineae." *Proc. Natl. Acad. Sci. USA* **94**, 6861.

Barry, G. (2001) "The use of the Monsanto draft rice genome sequence in research." *Plant Phys.* **125**, 1164-1165.

Bentley, D. *et al.* (2001) "The physical maps for sequencing human chromosomes 1, 6, 9, 10, 13, 20 and X." *Nature* **409**, 942-943.

Chen, M., Presting, G., Barbazuk, W., Goicoechea, J., Blackmon, B., Fang, G., Kim, H., Frisch, D., Yu, Y., Higingbottom, S., Phimphilai, J., Phimphilai, D., Thurmond, S., Gaudette, B., Li, P., Liu, J., Hatfield, J., Sun, S., Farrar, K., Henderson, C., Barnett, L., Costa, R., Williams, B., Walser, S., Atkins, M., Hall, C., Bancroft, I., Salse, J., Regad, F., Mohapatra, T., Singh, N., Tyagi, A., Soderlund, C., Dean, R. and Wing, R. (2001) "An integrated physical and genetic map of the rice genome." *Plant Cell* **14**, 537-545.

Coulson, A., Sulston, J., Brenner, S. and Karn, J. (1986) "Towards a physical map of the genome of the nematode *C. elegans*." *Proc. Natl. Acad. Sci. USA* **83**, 7821-7825.

Devos, K. and Gale, M. (1997) "Comparative genetics in the grasses." *Plant Mol. Biol.* **35**, 3-15.

Green, P. (1997) "Against a whole-genome shotgun." *Genome Research* **7**, 410-417.

Gregory, S., Howell, G. and Bentley, D. (1997) "Genome mapping by fluorescent fingerprinting." *Genome Research* **7**, 1162-1168.

Hoskins, R., Nelson, C., Berman, B., Laverty, T., George, R., Ciesiolka, L., Naeemuddin, M., Arenson, A., Durbin, J., David, R., Tabor, P., Bailey, M., DeShazo, D., Catanese, J., Mammoser, A., Osoegawa, K., de. Jong, P., Celniker, S., Gibbs, R., Rubin, G. and Scherer, S. (2000) "A BAC-based physical map of the major autosomes of *Drosophila melanogaster.*" *Science* **287**, 2271-2274.

Marra, M., Kucaba, T., Dietrich, N., Green, E., Brownstein, B., Wilson, R., McDonald, K., Hillier, L., McPherson, J. and Waterston, R. (1997) "High throughput fingerprint analysis of large-insert clones." *Genome Research* **7**, 1072-1084.

Marra, M., Kucaba, T., Sakhon, M., Hillier, L., Martienssen, R., Chinwalla, A., Crockett, J., Fedele, J., Grover, H., Gund, C., McCombie, W., McDonald, K., McPherson, J., Mudd, N., Parnell, L., Schein, J., Seim, R., Shelby, P., Waterston, R. and Wilson, R. (1999) "A map for sequence analysis of the *Arabidopsis thaliana* genome." *Nature Genetics* **22**, 265-275.

Myers, E., Sutton, G., Delcher, A., Dew, I., Fasulo, D., Flanigan, M., Kravitz, S., Mobarry, C., Reinert, K., Remington, K., Anson, E., Bolanos, R., Chou, H., Jordan, C., Halpern, A., Lonardi, S., Beasley, E., Brandon, R., Chen, L., Dunn, P., Kai, K., Liang, Y.D., Nusskern, M., Zhang, Q., Zheng, X., Runbin, G., Adams, M. and Venter, J. (2000) *Science* **287**, 2196-2204.

Olson, M., Dutchik, J., Graham, M., Brodeur, G., Helms, C., Frank, M., MacCollin, M., Scheinman, R. and Frank, T. (1986) "Random-clone strategy for genomic restriction mapping in yeast." *Proc. Natl. Acad. Sci. USA* **83**, 7826-7830.

San Miguel, Tikhonov, A., Jin, Y., *et al.* (1996) "Nested retrotransposons in the intergeneic regions of the maize genome." *Science* **274**, 765-768.

Sasaki, T. and Burr, B. (2000) "International rice genome sequencing project: the effort to completely sequence the rice genome." *Curr. Opinion in Plant Biol.* **3**, 138-141.

Soderlund, C., Longden, I. and Mott, R. (1997) "FPC: a system for building contigs from restriction fingerprinted clones." *CABIOS* **13**, 523-535.

Soderlund, C., Humphrey, S., Dunhum, A. and French, L. (2000) "Contigs built with fingerprints, markers and FPC V4.7." *Genome Research* **10**, 1772-1787.

Sulston, J., Mallet, F., Staden, R., Durbin, R., Horsnell, T. and Coulson, A. (1988) "Software for genome mapping by fingerprinting techniques." *CABIOS* **4**, 125-132.

Sulston, J., Mallett, F., Durbin, R. and Horsnell, T. (1989) "Image analysis of restriction enzyme fingerprints autoradiograms." *CABIOS* **5**, 101-132.

The Arabidopsis Genone Initiative (2000) "Analysis of the genome sequence of the flowering plant *Arabidopsis thaliana.*" *Nature* **408**, 769-815.

The International Human Genome Mapping Consortium (2001) "A physical map of the human genome." *Nature* **409**, 934-941.

The International Human Sequencing Consortium (2001) "Initial sequencing and analysis of the human genome." *Nature* **409**, 860-920.

Venter, *et al.* (2001) "The sequence of the Human Genome." *Science* **291**, 1304-1351.

Venter, J.C., Smith, H.O. and Hood, L. (1996) "A new strategy for genome sequencing." *Nature* **381**, 364-366.

Vicient, C., Kalendar, R. and Anamthawat-Jonsson, K. (1999) "Structure, functionality, and evolution of the BARE-1 retrotransposon of barley." *Genetica* **107**, 53-63.

Ware, D., Jaiswal, P., Ni, J., Pan, X., Chang, K., Clark, K., Teytelman, L., Schmidt, S., Zhao, W., Cartinhour, S., McCouch, S., and Stein, L. (2002). "Gramene: a resource for comparative grass genomics." *Nucleic Acids Research* **30**, 103-105.

Weber, L. and Myers, E. (1997) "Human whole-genome shotgun sequencing." *Genome Research* **7**.

Wing, R.A., Yu, Y., Presting, G., Frisch, D., Wood, T., Woo, S-S., Budiman, M.A., Mao, L., Kim, H.R., Rambo, T., Fang, E., Blackmon, B., Goicoechea, J.L., Higingbottom, S., Sasinowski, M., Tomkins, J., Dean, R.A., Soderlund, C., McCombie, R., Martienssen, R., de la Bastide, M., Wilson, R. and Johnson, D. (2001) "Sequence-tagged connector/DNA fingerprint framework for rice genome sequencing." In Khush, G., Brar, D. and Hardy, B. (eds). Rice Genetics IV, International Rice Research Institute, Science Publishers.

Yuan, Q., Liang, F., Hsiao, J., Zismann, V., Benito, M., Quackenbush, J., Wing, R. and Buell, R. (2001) "Anchoring of rice BAC clones to the rice genetic map in silico." *Nucleic Acid Research* **28**, 3636-3641.

Authors' Addresses

Carol Soderlund, Clemson University, Genomic Institute, 100 Jordan Hall, Clemson, SC 29634-5808, USA. Current Address: Plant Science Department, 303 Forbes Building, University of Arizona, Tucson, AZ 85721, USA. Email: cari@genome.arizona.edu.

Fred Engler, Clemson University, Genomic Institute, 100 Jordan Hall, Clemson, SC 29634-5808, USA.

James Hatfield, Clemson University, Genomic Institute, 100 Jordan Hall, Clemson, SC 29634-5808, USA.

Steven Blundy, Clemson University, Genomic Institute, 100 Jordan Hall, Clemson, SC 29634-5808, USA.

Mingsheng Chen, Clemson University, Genomic Institute, 100 Jordan Hall, Clemson, SC 29634-5808, USA.

Yeisoo Yu, Clemson University, Genomic Institute, 100 Jordan Hall, Clemson, SC 29634-5808, USA.

Rod Wing, Clemson University, Genomic Institute, 100 Jordan Hall, Clemson, SC 29634-5808, USA.

Chapter 4

Graph Theoretic Sequence Clustering Algorithms and Their Applications to Genome Comparison

Sun Kim

4.1 Introduction

Recent advances in both sequencing technology and algorithmic development for genome sequence software have made it possible to determine the sequences of whole genomes. As a consequence, the number of completely sequenced genomes is increasing rapidly. In particular, as of December 2001, there are more than 65 completely sequenced genomes in the GenBank. However, algorithmic development for the genome annotation is relatively slow and annotation of the completely sequenced genome inevitably relies on human expert knowledge. Since the accurate annotation of genomic data is of supreme importance, human experts need to annotate the genomic data. This manual annotation process on a large amount of data is prone to errors. The quality of annotation can be significantly improved by using robust computational tools. One of the most important class of computational tools is the sequence clustering algorithm. Recently developed clustering algorithms [Tatusov *et al.*, 1997; Matsuda *et al.*, 1999; Enright and Ouzounis, 2000; Matsuda, 2000] were successful in clustering a large number of sequences simultaneously, e.g. whole sets of proteins from multiple organisms. In this chapter we review these algorithms briefly and present our sequence clustering algorithm BAG based on graph theory.

4.1.1 Database search as a genome annotation tool

Suppose that we annotate a set of sequences $S = \{s_1, s_2, ..., s_n\}$. The most widely used method is to search for each sequence s_i against the sequence databases. If there is a strong evidence in terms of statistical analysis value such as E-value, we may conclude that s_i belongs to a certain family F_j. Otherwise, we skip the annotation of s_i.

One of the main problems with the database search strategy is that the search result needs to be evaluated manually by human experts. This process requires too much human intervention, and the quality of annotation largely depends on the knowledge and work behavior of human experts. In addition, this process assumes that the database annotation and sequence classification are correct. If neither is correct, then annotation errors could propagate.

Issues with the database search

We limit our discussion to issues related to the sequence analysis: for example, the accuracy of the annotations in the database is not discussed here. Although the database search is probably the most widely used tool for the annotation of sequences, there are three main issues that users have to deal with.

The cutoff threshold setting issue

The database search with a query sequence returns matches that sometimes are not biologically related to the query, i.e. *false positives*. For this reason, each match is associated with a score that shows a statistical significance of the match. The statistical scores, such as Zscore and Evalue, are very effective in ranking the matches in relation to the biological significance, often termed as *homology*. Thus, biologists select a certain threshold score to filter out false positives and matches above the score are trusted as true positives. A strict cutoff threshold may result in discarding many matches that are homologous to the query sequences, i.e., too many false negatives, while a relaxed cutoff threshold may result in including many false positives. Due to this difficulty, a cutoff score is subjectively determined and biologists need to look at search results one by one even when there are many query sequences to be searched for. In addition, it is not possible to set an absolute threshold value for database searches with an arbitrary query sequence. In general this is an issue for any kind of biological sequence analysis.

The remote homology detection issue

There are cases where two sequences, s_1 and s_2, are not similar but share the same functions. In these cases, the database search with s_1 may miss the match s_2. This problem can be effectively addressed by including intermediate sequences [Park *et al.*, 1997]. We can utilize this fact to identify remote homology sequences by iteratively performing the search with strong sequence matches from the previous search (see Figure 1). Indeed, there are several database search algorithms that are very successful in detecting remote homologous sequences by automatically incorporating intermediate sequences. Among them is PSI-BLAST [Altschul *et al.*, 1997], which iteratively searches the database with a profile constructed from a previous database search. The issue here is to determine which sequences should be included as intermediate sequences with which the search can be iterated.

The transitivity bounding issue

Iterative searches with intermediate sequences can be viewed as building transitive relationships from sequence s_i to sequence s_k by chaining a relationship from s_i to s_j and another from s_j to s_k. Unfortunately, chaining can lead to false positives, so care must be taken when to terminate the chain. For example, the fourth iteration with s_4 in Figure 1 would result in identifying false positives.

Assuming that the geometry of sequence relationships is known, these three issues are illustrated in Figure 1. After the initial database search with **q**, we choose a sequence among the matches as the next query, i.e. ·an intermediate sequence. Which sequences should be the query for the next round iteration? Figure 1 shows the second round search with **s1**. Why not with **s2** or both? This is the selection of intermediate sequences in the remote homology detection issue. In this example, the search stops at the third iteration. Another iteration would result in adding mostly false positives. How many iteration would be appropriate? It obviously depends on the

Search with q

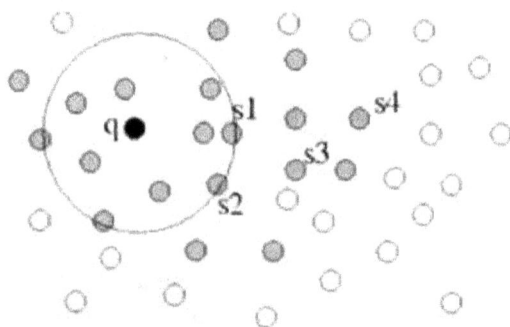

Search with s1

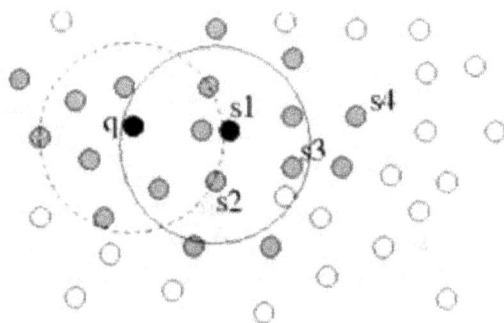

Search with s3

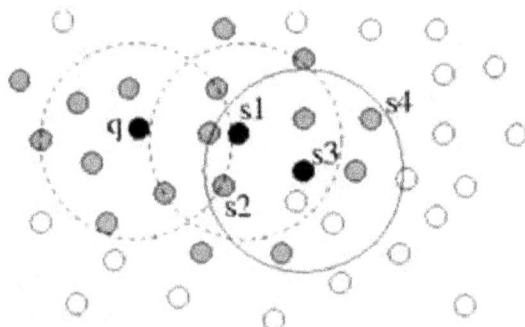

Figure 1. Illustration of database searches assuming that the geometry of sequence relationships is known. Small circles denote sequences: filled ones are of the same family and unfilled ones are not. Big circles with centers, q, s1, and s3, denote database searches with the sequences respectively. The figure represents a series of three iterated database searches with queries, q, s1, and then s3.

query, the database, and the sensitivity of the search tool, and there is no absolute answer. This is the transitivity bounding issue.

4.1.2 Clustering algorithms as a genome annotation tool

Clustering algorithms use structures of sequence relationships to classify a set of sequences into families of sequences, $F = \{F_1, F_2, ..., F_n\}$. While generating F, the remote homology detection issue and the transitivity bounding issue are systematically addressed with the structures of sequence relationships used by the clustering algorithm. Any two sequences, s_i and s_j, in the same family $F_i = \{..., s_i, s_j, s_k,..\}$ are related by intermediate sequences, say s_k, even when there is no observable relationship between s_i and s_j, thus the remote homology detection issue is addressed. The sequences s_i and s_m in two different families could be related through intermediate sequences $s_{m1}, ...,$ s_{mi} but such chaining of sequence relationships is blocked if the structure of sequence relationships used in the clustering algorithm classifies s_i and s_m into two different families. Thus the transitivity issue is addressed. In addition, clustering algorithms simultaneously analyze all input sequences, not one by one. Thus there is only one analysis output, though it may contain a large amount of information, which needs to be verified by human experts.

In addition, clustering algorithms generally use only sequence information, not annotations. Consequently results from clustering are not sensitive to potential errors in annotations, which may be used for verification of previous annotations in the database.

Recently developed sequence clustering algorithms were successful in clustering a large number of sequences into sequence families of highly specific categories. These clustering algorithms used graph theory explicitly or implicitly. The next section will briefly summarize these clustering algorithms. We will then discuss our graph theoretic sequence clustering algorithm.

4.2 Preliminaries

In this section, terminolgies on graphs are introduced. The definitions are drawn from [Cormen, 1989].

A *graph G* is a pair (V,E), where V is a finite set and E is a binary relation on V. The set V is called the *vertex set* of G; its elements are called *vertices*.

The set E is called the *edge set* of G, and its elements are called *edges*. An edge from u to v is denoted as (u,v). If (u,v) is an edge in a graph, vertex u is *adjacent* to vertext v. The edge (u,v) is incident to vertex u and vertex v. The degree of a vertex v, denoted by $deg(v)$, is the number of edges incident to v. A *path* of *length* k from a vertex u to a vertex u' is a sequence $<v_0, v_1,..., v_k>$ of vertices such that $u = v_0$, $u' = v_k$, and (v_{i-1}, v_i) is in E for $i = 1,2,...,k$. If there is a path p from u to u', u' is *reachable* from u via p. A path is *simple* if all vertices in the path are distinct, i.e., $v_i \neq v_j$ for $0 \leq i, j \leq k$.

A graph $G' = (V',E')$ is a subgraph of $G = (V,E)$ if $V' \subseteq V$ and $E' \subseteq E$. A subgraph of G *induced* by V' is the graph $G' = (V',E')$, where $E' = \{(u,v) \in E: u,v \in V'\}$. In a directed graph, a path $< v_0, v_1,..., v_k>$ is said to be a *cycle* if $v_0 = v_k$ and $k \neq 0$. A cycle is *simple* if all vertices are distinct.

An undirected graph is *connected* if there is a path for every pair of vertices. The *connected component* of a graph is a subgraph where any two vertices in the subgraph are reachable from each other. An *articulation point* of G is a vertex whose removal disconnects G. For example, in Figure 2 the removal of a vertex $s5$ disconnects G. A *biconnected graph* is a graph where there are at lest two disjoint paths for any pair of vertices. A *biconnected component* of G is the maximal biconnected subgraph. In Figure 2, a subgraph G_1 induced by vertices $\{s2, s3, s4\}$ is a biconnected graph but it is not maximal since another subgraph G_2 induced by vertices $\{s1, s2, s3, s4, s5\}$ is biconnected and G_1 is a subgraph of G_2. There are two biconnected components, $\{s1, s2, s3, s4, s5\}$ and $\{s5, s6, s7, s8, s9\}$ of G.

A *complete graph* is an undirected graph in which every pair of vertices are adjacent. A *hypergraph* is an undirected graph where each edge connects arbitrary subsets of vertices rather than two vertices.

4.3 Sequence Clustering Algorithms Based on Graph Theory

Given a set of biological sequences, $S = \{s_1, s_2, ..., s_n\}$, pairwise sequence relationships can be established using pairwise sequence alignment algorithms such as FASTA [Pearson *et al.*, 1988], BLAST and PSI-BLAST [Altschul *et al.*, 1990; Altschul *et al.*, 1997], and Smith-Waterman algorithm [Smith and Waterman, 1981]. A pairwise relationship (s_i, s_j) can be thought as an edge between two vertices s_i and s_j in a graph. In this way, we can build a graph from a set of pairwise matches. The graph can then be used to build a structure among sequence relationships, which represents families of sequences.

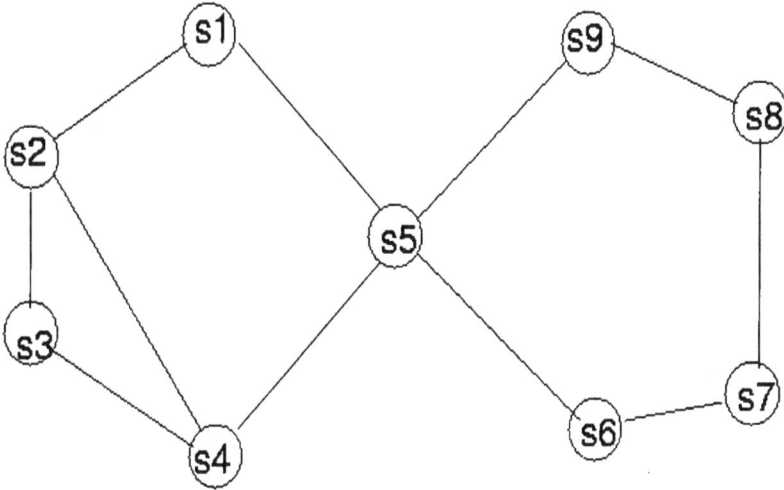

Figure 2. Biconnected components and articulation points. The vertex s5 is an articulation point since removing the vertex results in separating the graph.

In this section, we summarize recent developments in sequence clustering algorithms that were successful in clustering a large number of sequences, e.g., whole sets of predicted proteins from multiple genomes, into families of specific categories. The general approach can be summarized as below.

1. Compute similarities for every pair of sequences.
2. Build a sequence graph G from the pairwise matches above a preset cutoff threshold.
3. Generate a set of subgraphs { G_1, G_2, ..., G_n } with respect to certain graph structures employed in the clustering algorithm.
4. The set of vertices in each subgraph G_i forms a family of sequences.

We will call G the *sequence graph*. Graph based clustering algorithms need to handle multidomain sequences as they generate a set of families. Multidomain sequences belong to multiple families, thus the vertex sets of subgraphs are not disjoint. However, the edge sets of subgraphs are expected to be disjoint since a multidomain sequence has multiple edges, probably as many as the number of domains in the sequence, to other sequences. In this sense, the problem we are discussing is the clustering problem for pairwise sequence relationships rather than for sequences themselves. However, all sequence clustering algorithms generate a set of families, each of which is a set of sequences. Thus the graph problem we are dealing with is a *graph covering problem*, where subgraphs can share the same vertices, rather than a *graph partitioning problem*, where no two subgraphs share the same vertices.

This section summarizes four sequence clustering algorithms based on graph theory for the sequence clustering problem [Tatusov *et al.*, 1997; Matsuda *et al.*, 1999; Enright and Ouzounis, 2000; Matsuda, 2000]. Although GeneRAGE [Enright and Ouzounis, 2000] does not explicitly use specific graph properties for clustering sequences, we will provide our interpretation of the algorithm in terms of graph theory.

4.3.1 Matsuda, Ishihara, and Hashimoto algorithm

The Matsuda, Ishihara, and Hashimoto algorithm [Matsuda *et al.*, 1999] introduced a graph structure called *p-quasi complete graph* for describing a family of sequences with a confidence measure and used the graph structure to generate a set of subgraphs from the sequence graph.

A set of sequences, $S = \{s_1, s_2, ..., s_n\}$, that belong to the same family, will have at least one conserved domain, and subsequences corresponding to the same domain will share certain levels of sequence similarity. If all subsequences are highly similar, every pairwise sequence relationship (s_i, s_j), can be detected and the resulting graph will be a *complete graph*, i.e., every pair of vertices is connected. However, some sequences in the same family may be distant in terms of similarity, and we cannot expect a sequence graph from the sequences in the same family to be complete. Thus, a graph property called *p-quasi completeness* was proposed. A graph is a p-quasi complete graph if $deg(v) \geq p$ for all $v \in V$.

Multidomain proteins are handled by formulating the sequence clustering problem as a *graph covering problem*, where a multidomain sequence may belong to more than one subgraphs, i.e., covers. The problem is then to search for the *minimum* number of covers of the given sequences such that each cover is represented as a *maximal p*-quasi complete subgraph. Matsuda *et al.* [1999] proved that the *p*-quasi complete subgraph covering problem is NP-complete. Thus a heuristic algorithm was proposed.

In an experiment of clustering 4,586 proteins from *E. coli*, they were able to classify the protein set into 2,507 families with $p = 0.4$ and 2,747 families with $p = 0.8$ with a SW score cutoff value of 100. They also reported discovering multidomain proteins.

This clustering algorithm used p-quasi completeness of a subgraph as a clustering confidence measure. Thus the remote homology detection and transitivity bounding issues are handled with p-quasi completeness. However, the cutoff threshold setting issue still remains unresolved since users need to set the value without guidance. In addition, setting the connectivity ratio remains unresolved as discussed in the paper. The complexity of the algorithm is $O(n^4)$ where n is the number of sequences.

4.3.2 Matsuda algorithm

Matsuda [2000] proposed an algorithm for detecting conserved domains in a set of protein sequences using the *density* of a graph for clustering measure. The algorithm works as below.

1. All possible subsequences (blocks) of length l are generated from a set of protein sequences and all pairwise comparisons of blocks are computed.
2. A *block graph* is computed using a set of pairwise matches whose scores are greater than a preset cutoff value.
3. A set of maximum density subgraphs of the block graph are computed.
4. Overlapping blocks are combined into larger blocks.

The density of a graph G_i is defined as an average weighted sum of all edges in G_i, where the weight of an edge is defined as the similarity score. The maximum density subgraph G_i^* of G_i is a subgraph whose average weighted sum of its edges is the maximum among all possible subgraphs. This can be computed in $O(|V|^3)$ where $|V|$ is the number of vertices. This algorithm was able to detect multidomain proteins in experiments with a set of transcription regulation proteins, a set of *E. coli* two component system proteins, and a set of δ^{70} factor proteins.

This clustering algorithm used the density as a clustering confidence measure. Thus the remote homology detection and transitivity bounding issues are handled with the maximum density of a subgraph that does not require user input parameters, in contrast to the value of p in the p-quasi completeness in [Matsuda *et al.*, 1999]. The cutoff threshold setting is still an issue, but a sufficiently low cutoff value can be used for the clustering analysis. The complexity of the proposed algorithm is $O(\|P\|^3)$ where $\|P\|$ denotes the total sum of input protein sequences.

4.3.3 COG: cluster of orthologous groups

Tatusov *et al.* [1997] proposed an effective method for clustering proteins from completely sequenced genomes and constructed a database of protein families called COGs (*Cluster of Orthologous Groups*) [Tatusov *et al.*, 2001].

The COG database was created using the concept of orthologous and paralogous genes. *Orthologs* are genes in different species that evolved from a common ancestral gene by speciation. *Paralogs* are genes related by duplication within a genome. Orthologs are modeled as a genome context best hits (BeTs) and extended later by clustering analysis. The outline for COG database construction is as follows.

1. Perform all pairwise protein sequence comparisons.
2. Detect and collapse obvious paralogs, that is, proteins from the same genome that are more similar to each other than to any proteins from other species.
3. Detect triangles of mutually consistent, genome-specific best hits (BeTs), taking into account the paralogous groups detected at the above step.
4. Merge triangles with a common side to form COGs.
5. A case-by-case analysis of each COG is performed. This analysis serves to eliminate false-positives and to identify groups that contain multidomain proteins by examining the pictorial representation of the BLAST search outputs. Multidomain proteins are split into single-domain segments and the steps 2 - 5 are repeated.
6. Examine large COGs visually with phylogenetic analysis and split them into small clusters if needed.

The use of BeTs, the best hit in the context of genomes, for establishing sequence relationships handles the cutoff threshold setting issue: once a sufficiently low cutoff value is set, matches through BeT can be trusted as true positives. The requirement of a triangle of BeTs and subsequent merging of the trianlgles becomes an effective way of clustering sequences. From a graph theoretical perspective, the BeT graph is a directed graph and the resulting cluster requires *at least* being a strongly connected component. It would be interesting to study the characteristics of the graph structure used in the COG database construction.

As a result of using the strict sequence relationship, i.e., BeT, and the graph structure, the COG database was successful clustering proteins from many completed genomes into families of very specific categories. As of December 2001, there are 3,311 COGs from 44 complete genomes.

4.3.4 GeneRAGE

Enright and Ouzounis [2000] proposed a sequence clustering algorithm based on single linkage clustering after the symmetrification and the multidomain protein detection steps. The outline of the algorithm is as follows.

1. Given a set of sequences, $S=\{s_1, s_2,...,s_n\}$, the pairwise sequence relationships are computed using BLAST 2.0 [Altschul *et al.*, 1997] after masking compositionally biased regions. Any pairwise match (s_i, s_j) with Evalue $< 10^{-10}$ is accepted and recorded as $T(s_i, s_j) = 1$ in a matrix T.

2. The symmetrification step enforces a symmetry of a pairwise relationship: $T(s_i, s_j) = 1$ if and only if $T(s_j, s_i) = 1$. Note that the pairwise comparison is not symmetric. For any pair $T(s_j, s_i) \neq T(s_i, s_j)$, a more rigorous sequence alignment with SW algorithm [Smith and Waterman, 1981] is performed. If the Zscore is greater than 10, both $T(s_j, s_i)$ and $T(s_i, s_j)$ are set to 1. Otherwise, both $T(s_j, s_i)$ and $T(s_i, s_j)$ are set to 0.

3. The multidomain protein detection step checks for a cycle among three proteins. If $T(s_i, s_k) = 1$ and $T(s_j, s_k) = 1$, i.e., s_i and s_j are matched to the same sequence s_k, a symmetric pairwise relationship between s_i and s_j is enforced. If the symmetric relationship does not hold, s_k becomes a candidate for a multidomain protein. Note that both $T(s_i, s_j)$ and $T(s_j, s_i)$ are set to 0 if s_k becomes a multidomain candidate.

4.   Single linkage clustering is iteratively performed for any sequence that is not already in the cluster.

The algorithm clustered the set of proteins from *M. jannaschii* into 61 families. Among them, 58 families were consistent with manual annotation while 3 families have conflicting annotations.   The multiple sequence alignments of 3 families were consistent, indicating possible incorrect annotations. (The multiple sequence alignments were not shown in their paper.) The algorithm was successful in finding muldomain proteins in the genome. The paper also reports the discovery of 294 new families in the PFAM [Bateman *et al.*, 2000] database release 5.

The multidomain detection step refines the sequence relationship to have a triangle relationship for every protein. Thus, from a graph theoretic perspective, their clustering algorithm does require a graph structure similar to the COG database construction algorithm.

4.4 A New Graph Theoretic Sequence Clustering Algorithm

We present a new graph theoretic sequence clustering algorithm that explicitly uses two graph properties: biconnected components and articulation points (see Figure 2). A biconnected component corresponds to a family of sequences and an articulation point corresponds to a multidomian protein. Since an articulation point is the only vertex that connects multiple biconnected components, i.e., multiple families, it is intuitive to consider each articulation point as a candidate for multidomain sequence.

4.4.1 The basic algorithm

A simple version of our algorithm follows the general procedure described in Section 4.3.

Given a set of sequences $S = \{s_1, s_2, ..., s_n\}$,

1.   Compute similarities (s_i, s_j) for all $1 \leq i, j \leq n$ and $i \neq j$.
2.   Build a sequence graph G from the pairwise matches above a preset cutoff threshold. Generate a set of subgraphs, $\{G_1, G_2, ..., G_m\}$, each of which G_i is a biconnected component.

3. Then a set of vertices in each subgraph G_i forms a family of sequences and each articulation point becomes a candidate for multidomain sequence.

To reduce the computation time in Step 1, we can use well accepted approximation algorithms such as FASTA [Pearson *et al.*, 1988] or BLAST [Altschul *et al.*, 1990; Altschul *et al.*, 1997]. We simply choose FASTA for the pairwise computation, and so the computation is FASTA (s_i, S) for all $1 \leq i \leq n$.

The complexity of the algorithm is $O(n^2)$ for n sequences. Step 1 requires $n \times (n-1)$ pairwise comparisons and Step 3, the computation of biconnected components, is proportional to the number of edges in the graph, i.e., the number of pairwise matches above a preset cutoff threshold. However, the algorithm runs much faster in practice. Use of the FASTA algorithm for pairwise matches requires n searches against S and each FASTA search with s_i, FASTA(s_i, S), runs much faster than $(n-1)$ pairwise sequence comparisons between s_i and s_j for $1 \leq j \leq n$ and $i \neq j$ since FASTA is an approximation algorithm with hashing techniques. The number of edges is much smaller when we discard pairwise matches below the cutoff threshold. In cases where the clustering analysis can be done on precomputed pairwise databases such as [Cannarozzi, 2000; Gilbert, 2002] the complexity of our algorithm becomes linear in relation to the number of pairwise matches above a preset cutoff threshold. This computational efficiency becomes an critical feature for an extended version of the algorithm and addresses the cutoff threshold setting issue discussed in Section 4.1.1.

4.4.2 Result from application of the basic algorithm

We performed a clustering analysis of all 1,881 predicted protein sequences from *Borrelia burgdorferifull* (GENBANK accession number AE000783, 850 proteins) and *Treponema pallidum* (GENBANK accession number AE000520, 1031 proteins) with the Zscore cutoff of 200. 470 families of 1,076 sequences are clustered with 42 multidomian candidates, excluding families that contain a single sequence (these families are uninformative). Most of the resulting families are clustered correctly with high precision according to the current annotation. For example, 102 ribosomal proteins in the two genomes are grouped into 51 families of 2 proteins according to subunits matching in the two genomes except S1 subunit proteins. The S1 subunit proteins were an interesting case with three families shown below. The sequence graph is shown in Figure 3.

Family 133

Articulation point: gil3322552 to families: 133 134 135
>gil3322552lgblAAC65266.1l ribosomal protein S1 (rpsA) [*Treponema pallidum*]
>gil2688007lgblAAC66509.1l cytidylate kinase (cmk-1) [*Borrelia burgdorferi*]

Family 134

Articulation point: gil3322552 to families: 133 134 135
>gil3322552lgblAAC65266.1l ribosomal protein S1 (rpsA) [*Treponema pallidum*]
>gil3323244lgblAAC65881.1l tex protein (tex) [*Treponema pallidum*]

Family 135

Articulation point: gil3322552 to families: 133 134 135
>gil3322552lgblAAC65266.1l ribosomal protein S1 (rpsA) [*Treponema pallidum*]
>gil2688008lgblAAC66510.1l ribosomal protein S1 (rpsA) [*Borrelia burgdorferi*]

From the annotations in the heading, it was not obvious that Families 133 and 134 were clustered correctly. Thus, we performed protein domain search against PFAM [Bateman *et al.*, 2000] at pfam.wustl.edu[1] and Family 133 was confirmed to share Cytidylate kinase domain and Family 134 was confirmed to share S1 RNA binding domain as shown in Table 1. Indeed, the sequence gi3322552, which was an articulation point, had both domains (see Table 1). The only concern was how to merge Families 134 and 135 into one. If we relax the cutoff threshold to Zscore of 150, the two families become one, i.e., single connected components as shown in Figure 4. However, the question is *how do we know the cutoff threshold value 150 a priori?* This is the cutoff threshold setting issue discussed in Section 4.1.1! This fundamental issue will be effectively addressed in the extended version of our algorithm described in the following section.

---

[1] The PFAM version we used is 6.6.

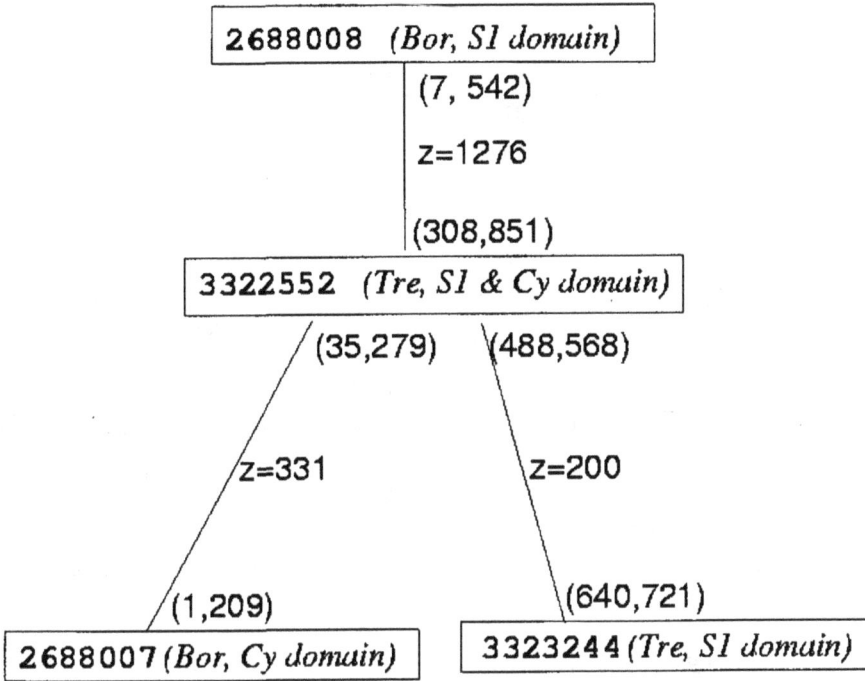

Figure 3. A sequence graph with the Zscore cutoff threshold of 200. The numbers in parantheses denote the intervals of the overlapping regions.

4.5 BAG: The Extended Clustering Algorithm

The basic algorithm in Section 4.4.1 is extended and called Biconnected components and Articulation points based Grouping of sequences (BAG).

4.5.1 Issues with the basic algorithm

There are several features of the basic algorithm presented in the previous section that need attention:

Sequence ID	from	to	Evalue	Domain
gi3322552	129	281	1.40E-83	Cytidylate kinase
	218	354	1.30E-07	S1 RNA binding domain
	395	432	0.0012	S1 RNA binding domain
	490	562	2.10E-21	S1 RNA binding domain
	575	649	2.50E-23	S1 RNA binding domain
	662	736	2.40E-17	S1 RNA binding domain
	749	825	1.40E-14	S1 RNA binding domain
gi3323244	642	715	2.80E-22	S1 RNA binding domain
gi2688007	59	211	2.30E-77	Cytidylate kinase
gi2688008	19	83	5.50E-09	S1 RNA binding domain
	184	257	3.40E-15	S1 RNA binding domain
	270	344	4.30E-21	S1 RNA binding domain
	357	430	2.40E-19	S1 RNA binding domain
	443	518	2.70E-08	S1 RNA binding domain

Table 1. The PFAM search result for four proteins in Families 133, 134, and 135.

1. **The cutoff threshold setting issue**: as shown with the example of the S1 and Cytidylate domain proteins, we do not know the cutoff threshold *a priori*, Zscore 150 for the example.
2. **Merging families**: Given a cutoff threshold, several families may need to be tested for merging as shown in the previous section.
3. **Spitting a family**: Given a cutoff threshold, a family may need to be tested for splitting into several ones.
4. **Multidomain protein**: How do we know an articulation point truly corresponds to a multidomain protein?

Each of these features will be explored in the following subsections.

4.5.2 Setting the cutoff threshold

We performed a series of clustering analyses with Zscore cutoff thresholds ranging from 100 to 1000 at 50 increment intervals for three sets

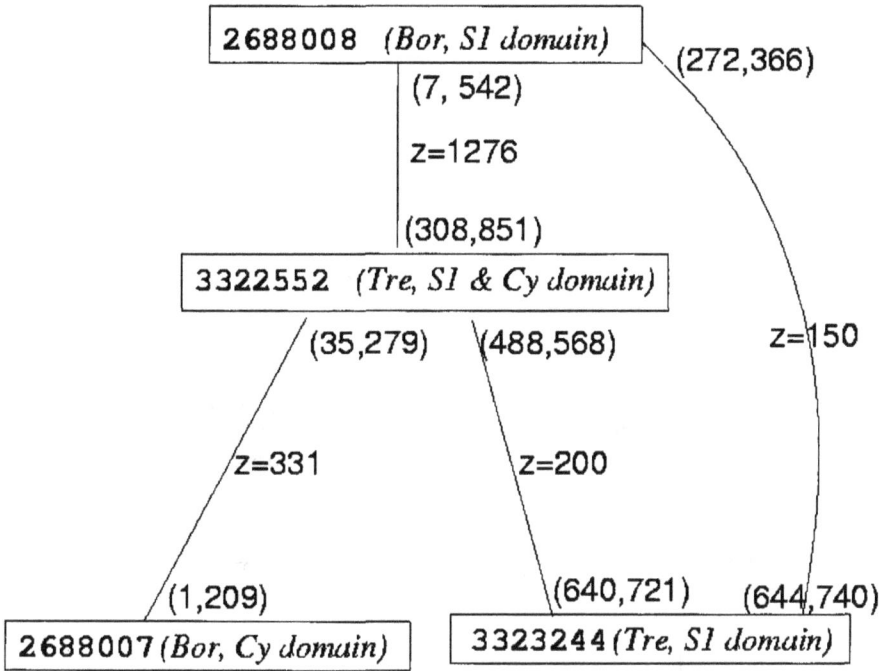

Figure 4. A sequence graph with the Zscore cutoff threshold of 150. The numbers in parantheses denote the intervals of the overlapping regions.

of pairwise comparisons from *B. burgdorferis* only, *T. pallidum* only and both genomes. Figure 5 plots the distributions of the number of biconnected components vs. Zscore cutoff thresholds and vs. SW score cutoff thresholds. We performed the two experiments to observe whether the resulting distribution would be significantly different for different scoring methods. Zscore is a statistical score that measures the likelihood of matches occurring by chance for a given database and its value depends on the size of the database. SW score is a sum of character match scores, and gap penalties and its value does not depend on the size of the database. The higher the score, the more significant a match is.

As we can observe in the figure, the number of biconnected components increases up to a certain value, 150 for Zscore and 100 for SW score, and then continues to decrease. The increase in the number of biconnected

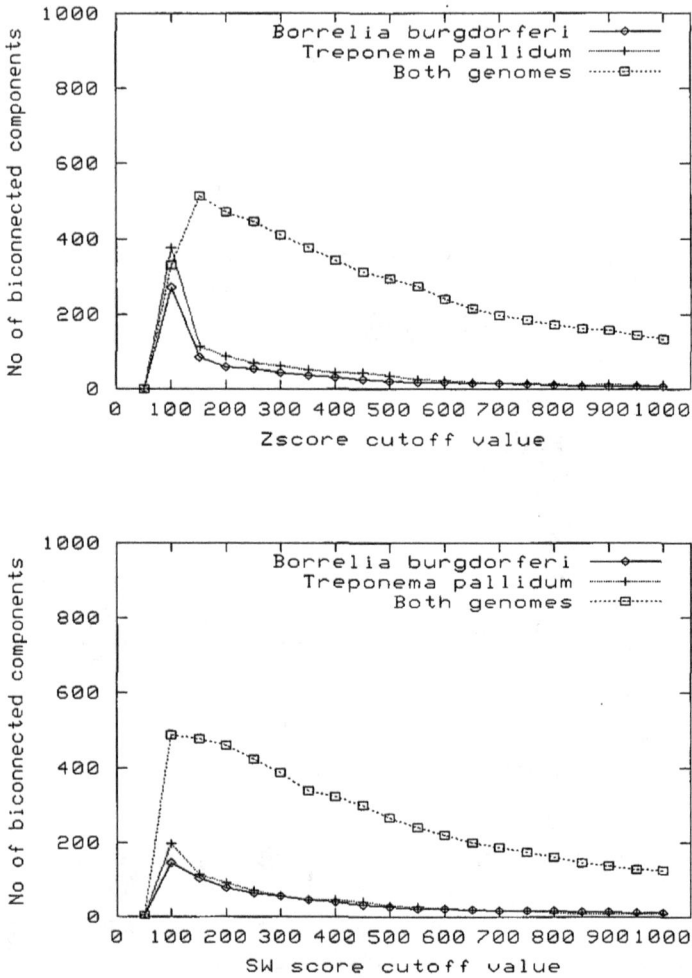

Figure 5. The distributions of the number of biconnected components vs. the Zscore cutoff thresholds (top plot) and SW scores (bottom plot).

components is intuitive, as a higher cutoff value will remove more false positives, thereby families of large size due to false positives being separated into several families. The decrease in the number of biconnected components is also intuitive as a higher cutoff value will remove more true positives, thereby more vertices become singletons, i.e., vertices without incident edges; note that singltons are not counted. We would expect that there exists a peak in the plot of the number of biconnected components vs. a score if the scoring method effectively models the pairwise sequence relationship. Zscore and SW score are well accepted scoring methods and have been verified empirically over the years in bioinformatics community. In this chapter, Zscore is used for the pairwise match score unless specified. Thus, by default, a *stricter* score means an increase in the score value, and a *relaxed* score means a decrease in the score value.

Note that the basic clustering algorithm runs in linear time in relation to the number of pairwise matches above a preset cutoff threshold after computing pairwise matches from a set of sequences. The series of clustering analysis with Zscore in Figure 5 took only 27 seconds on a Pentium IV 1.7 GHz processor machine running Linux. This computational efficiency makes it possible to effeciently conduct the series of clustering analyses with varying cutoff thresholds to find the cutoff threshold, $C_{maxbiconn}$, that generates the maximum number of biconnected components. However, we need to consider the number of articulation points as articulation points are candidates for multidomain proteins. Figure 6 shows the number of

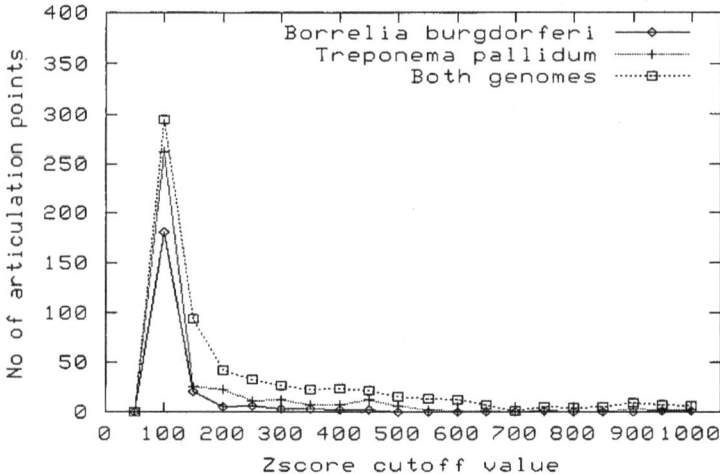

Figure 6. The distribution of the number of articulation points vs. Zscore cutoff thresholds.

articulation points with respect to varying Zscore cutoff thresholds. The articulation points become candidates for multidomain proteins and need to be tested for having multidomain proteins: the test method will be described in the following sections. Thus, we would avoid to select the cutoff threshold with too many articulation points. Let NAP_C be the number of articulation points at score C. One way to select the cutoff value is to use a ratio:

$$r = \frac{NAP_c}{NAP_{(c-I)}}$$

where I is the interval of the score for the series of clustering analysis.

4.5.3 The overview of the extended algorithm

1.   Build a graph G from the all pairwise comparisons result.
2.   Run the basic algorithm with cutoff scores ranging from C_1 to C_2 at each interval I and select a score, $C_{maxbiconn}$, where the number of biconnected components is the largest, and another score, C_{arti}, where the number of articulation points begins to decrease at a ratio $r < \Delta$.
3.   Select a cutoff score C_{arti} and generate biconnected components, G_1, G_2, ..., G_n with a set of articulation points $\{A_1, A_2, ..., A_m\}$
4.   Iteratively split a biconnected component into several ones with more stringent cutoff scores until there is no candidate component for splitting.
5.   Iteratively merge a set of biconnected components into one with relaxing the cutoff score to $C_{maxbiconn}$ until there is no candidate component for merging.

The overall procedure can be summarized in two steps: (1) generation of candidate families and (2) refinement of the families by merging and splitting. The fundamental question is which biconnected components need to be refined. For this purpose, we propose two tests as below.

1.   **AP-TEST** tests an articulation point for having potential multidomains.
2.   **RANGE-TEST** tests each biconnected component for being a single family.

Depending on the test result, splitting and merging operations are performed in a greedy fashion, i.e., once a subgraph is split or merged, it is not reconsidered for alternative splitting or merging. We will describe each test in detail.

To explain the details of this test, we introduce definition of notations.

Definition 1 MS(b_i, e_i) *denotes the match score between b_i and e_i, and refers to the Zscore of a FASTA search unless specified.*

Definition 2 *Given a set of sequences, $S=\{s_1, s_2,...,s_n\}$, and its sequence graph G whose edges are defined by pairwise matches of S,* **ALIGN**(s_i, s_j) *denotes a set of a pair of intervals, { [(b_{i1}, e_{i1}), (b_{j1}, e_{j1})], [(b_{i2}, e_{i2}), (b_{j2}, e_{j2})], ..., [(b_{ik}, e_{ik}), (b_{jk}, e_{jk})]} where each pair of intervals, [(b_{il}, e_{il}), (b_{jl}, e_{jl})] denotes the aligned regions between two sequences, (b_{il}, e_{il}) for s_i and (b_{jl}, e_{jl}) for s_j.*

Definition 3 *Given a pair of intervals $P=[(b_{i1}, e_{i1}), (b_{j1}, e_{j1})]$,* **INTERVAL1(P)** *denotes the first interval, i.e., (b_{i1}, e_{i1}) and* **INTERVAL2(P)** *denotes the second interval, i.e., (b_{j1}, e_{j1}).* **LENGTH**((b_i, e_i)) *denotes the length of the interval and is defined by $e_i - b_i +1$ if ($b_i \leq e_i$), 0 otherwise.*

Definition 4 *Given a pair of intervals $P= [(b_{i1}, e_{i1}), (b_{j1}, e_{j1})]$ and an interval $I=(I_1, I_2)$,* **PINTERSECT(I, P)** *returns an interval (b'_{j1}, e'_{j1}), where $b'_{j1} = b_{j1} + (b_{intersect} - b_{i1})$ and $e'_{j1} = e_{j1} - (e_{i1} - e_{intersect})$ and $b_{intersect} = MAX(b_{i1}, I_1)$ and $e_{intersect} = MIN(e_{i1}, I_2)$.*

PINTERSECT(I, P) computes an INTERVAL2(P) adjusted according to the intersection of I and INTERVAL1(P).

4.5.4 AP-TEST

The purpose of this test is to check for consistency in pairwise sequence overlapping regions at an articulation point.

We will illustrate AP-TEST using the sequence graph in Figure 3. By comparing two pairwise sequence overlaps in the sequence graph, ALIGN(gi3322552, gi2688007) = [(25,279), (1,209)], and ALIGN(gi3322552, gi3323244) = [(488,568), (640,721)], we find that two aligned regions in gi3322552, i.e., (25,279) and (488,568), do not overlap,

which is evidence that there might be two different functional domains in
gi3322552. Given an articulation point A_i, this procedure can be performed
with all vertices $v_{i1}, v_{i2}, ..., v_{in}$ adjacent to A_i. Since an articulation point is
expected to have multiple domains, its intersecting regions with adjacent
vertices should not share the same interval, i.e., multiple non-overlapping
intervals. The test AP-TEST(A_i) succeeds when there is no overlapping
interval for all adjacent vertices and fails when there is an overlapping
interval for all adjacent vertices.

The procedure AP-TEST(A_i) can be performed as below.

bool AP-TEST (vertex v)

```
1   i = (0,MAXINT)
2   for each vertex w adjacent to v do
3       i = INTERSECT(i, INTERVAL1(ALIGN(v,w)) )
4   done
5   if LENGTH(i) < Δ
6       then return true
7       else return false
8   endif
```

4.5.5 RANGE-TEST

Given a subgraph G_i induced by $\{v_1, v_2, ..., v_n\}$, we hope to test if all
overlapping regions ALIGN(v_j, v_k) share common intervals. If all sequences
in G_i share the same domain, it is expected that the intersection of all
overlapping regions will be greater than a certain length. To perform
RANGE-TEST(G_i), we need to order all vertices. One way to generate such
an order is to generate a Hamiltonian path where every vertex in G_i is visited
exactly once. The Hamiltonian path problem is known to be NP. Fortunately,
we just need only a path, which includes all vertices but vertices can be
visited more than once, for overlapping range checking purpose. Since every
subgraph is biconnected, we can easily compute such a path; we skip the
details on how to compute such a path in this chapter.

Once a path $p = < v_{i1}, v_{i2}, ..., v_{im} >$ is computed, we can check the
intersections of overlapping regions using PINTERVAL(I,P). For example,
consider a subgraph induced by a vertex set {gi3322552, gi3323244,
gi2688008} in Figure 4. Given a path $p = <gi3322552, gi3323244,$

gi2688008>, the region shared among the three sequences can easily be computed by chaining two pairwise overlaps, i.e., (gi3322552, gi3323244) and (gi3323244, gi2688008) as shown in Figure 7. The algorithm, RANGE-TEST(G_k), is described below.

As RANGE-TEST checks for consistency in overlapping regions among all sequences in the subgraph, the test RANGE-TEST(A_i) succeeds when there is an overlapping interval for all vertices, and fails when there is no overlapping interval for all vertices.

bool RANGE-TEST (G_k)

```
1   Compute a path p=< vi1, vi2, ..., vin > of Gk.
2      i = (0,MAXINT)
3   for j = 1 to (n-1) do
4            i = PINTERSECT(i, INTERVAL1(ALIGN(vij , vij+1 )))
5   done
6   if LENGTH(i) < Δ
7            then return false
8            else return true
9   endif
```

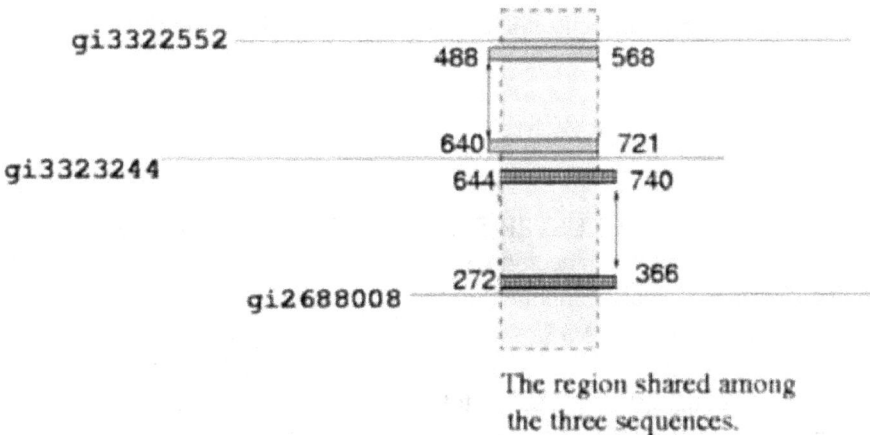

The region shared among
the three sequences.

Figure 7. The region shared among gi3322552, gi3323244, and gi2688008 can be computed by chaining two pairwise overlaps, i.e., (gi3322552, gi3323244) and (gi3323244, gi2688008).

4.5.6 HYPERGRAPH-MERGE

HYPERGRAPH-MERGE tests for merging multiple biconnected components by connectivity through articulation points.

Definition 5 *Given a set of biconnected components $\{G_1, G_2, ..., G_n\}$ and a set of articulation points $\{A_1, A_2, ..., A_m\}$, a* **hyper sequence graph H** *is a graph where vertices are G_i's and an edge (G_i, G_j) is defined when there is an articulation point A_k between G_i and G_j.*

The basic idea is to test if families, $G_{i1}, G_{i2}, ..., G_{im}$, can be merged into one. The candidate set for merging is determined again by computing biconnected components on the hyper sequence graph H. The algorithm for HYPERGRAPH-MERGE is given on the next page.

Merging subgraphs in *SG* is a greedy procedure. In line 4, the condition requiring that any of adjacent vertices is not yet merged is necessary since each merging on the hypergraph results in merging a set of sequences, not a single sequence. For example, suppose there are two biconnected components, $H_i = \{G_1, G_2\}$ and $H_j = \{G_2, G_3\}$, of H and both are merged successfully in line 8. Then *SG* contains two new subgraphs, G_i' from H_i and G_j' from H_j. Now all sequences in G_2 will belong to two different families, i.e., G_i' and G_j', which is not correct unless all the sequences in G_2 are multidomain sequences.

4.5.7 The Algorithm

The procedure CLUSTER-SPLIT(G, $C_{current}$, I) described below iteratively computes biconnected components of G with the score $C_{current}$ and splits – refines – each component with a stricter score $C_{current} + I$.

The algorithm for BAG described below is simply a three step process: (1) select two cutoff scores, C_{arti} and $C_{maxbiconn}$, and compute a set of biconnected components at the cutoff score C_{arti}, (2) iteratively split each biconnected component, and (3) iteratively merge several biconnected components into one.

bool HYPERGRAPH-MERGE ($C_{current}$, I)

SG: a global variable for the sets of biconnected components.
SA: a global variable for the sets of articulation points.
$C_{current}$: the current cutoff score.
I: incremental score.

1 **bool** *Merged* = false
2 Build a hyper sequence graph H with SG and SA.
3 Compute biconnected components, $H = \{ H_1, H_2, ..., H_k\}$.
   //Try to merge families in each component with a relaxed cutoff.//
4 **for** each H_i such that any vertex H_j adjacent to H_i is not merged **do**
5   Let $G_{H1}, G_{H2}, ..., G_{Hm}$ be vertices in H_i.
6   Create G' by merging $G_{H1}, G_{H2}, ..., G_{Hm}$.
7   Add new edges, (s_{hi}, s_{hj}), to G'
                 where $C_{current} \le \mathrm{MS}(s_{hi}, s_{hj}) \le (C_{current}+I)$
8   **if** (RANGE-TEST(G') is true) **then**
9                 $SG = SG - \{ G_{H1}, G_{H2}, ..., G_{Hm} \}$
10                $SG = SG \cup G'$
11   **for** each vertex $v \in SA$ such that
                      v is an edge between G_{Hi} and G_{Hj}
                      for some $1 \le i, j \le m$ **do**
12                        $SA = SA - \{v\}$
13   **done**
14   *Merged* = true
15   Mark H_i as merged
16   **endif**
17 **done**
18 **return** *Merged*

4.6 Implementation

The current prototype was implemented using C++ and an algorithmic library LEDA [Mehlhorn and Naher, 1999] on a Redhat 7.1 Linux machine. Because LEDA is a commercial package, we plan to develop a free ware version that includes public graph libraries such as the Boost Library [Siek *et al.*, 2002].

CLUSTER-SPLIT (G, $C_{current}$, I)

    SG: a global variable for the sets of biconnected components.
    SA: a global variable for the sets of articulation points.
    $C_{current}$: the current cutoff score
    I: incremental score

1   Compute biconnected components, $F=\{G_1, G_2,.., G_n\}$ of G, and
        articulation points, $A=\{A_1, A_2,.., A_m\}$.
2 **for** each A_j **do**
3   **if** (AP-TEST(A_j) is true)
4   **then** $SA = SA \cup \{ A_j \}$
5 **done**
    // Splitting a cluster. //
6 **for** each $G_j \in F$ **do**
7   **if** (RANGE-TEST (G_j) is false) **then**
8   // As G_j is not a family, use stricter score to refine G_j. //
9        $F = F$ - $\{ G_j \}$ // G_j is refined by a function call below. //
10   CLUSTER-SPLIT(G_j, $C_{current}+I$, I)
11   **endif**
12 **done**
13 $SG = SG \cup F$

The implementation can be obtained by contacting the author at sunkim@bio.informatics.indiana.edu or sun.kim@acm.org .

4.7 Application to Genome Comparison

In this section, we will discuss the application of our clustering algorithms to clustering entire protein sequences from complete genomes. With the current prototype, we were able to compare many different sets of genomes. The comparison of 63 whole bacterial and archeal genomes from GENBANK is underway and will be reported in a separate paper. In this section, we describe a complete analysis of two bacterial genomes, *B. burgdorferis* and *Treponema*.

BAG(M,I)

      INPUT: a set of pairwise matches $M=(s_i,s_j)$ and score intervals,
           I_{split} and I_{merge}.
      OUTPUT: a set of families F and a set of multidomain sequences, A.

1   $SG = \phi$ //A global variable for the sets of biconnected components.//
2   $SA = \phi$ // A global variable for the sets of articulation points. //
3   Build a graph G by including pairwise matches (s_i, s_j)
     where $MS(s_i, s_j) \geq C_{arti}$.
4   Get two scores, C_{arti} and $C_{maxbiconn}$ as described in Section 4.5.2.
5   CLUSTER-SPLIT(G, C_{arti}, I_{split})
6   $C = C_{arti}$
7   **while** $((C \geq C_{maxbiconn})$ and (HYPERGRAPH-MERGE (C, I_{merge})
    is true))
8        $C = C - I_{merge}$
9   **endwhile**
10 Report each $G_i \in SG$ as a family
11 Report each $v \in SA$ as a multidomain protein

As shown in Figure 5, we know that the number of biconnected components is the maximum at Zscore of 150. To find the Zscore of the maximum components at a finer scale, we performed a series of 11 clustering analysis with Zscore varying from 100 to 200 at each interval of 10 as shown in Table 2. We picked 110 for $C_{maxbiconn}$ and 200 for C_{arti} (see Section 4.5.2) and the clustering analysis starts with Zscore 200, i.e., C_{arti}, which clusters 470 families with 42 articulation points. Now we go through the details of how these 470 candidate families are refined, i.e., splitting with stricter scores and merging with relaxed scores.

To verify the clustering result, we used two methods, PFAM search at pfam.wustl.edu and the multiple sequence alignment for the cases where there is no domain detected by PFAM search. When we list the domains confirmed by PFAM search, these are the domain hits marked !! which means they are above the Pfam gathering cutoffs (GA) and are very significant hits that we would've automatically included in the PFAM full alignment.[2]

---

[2] The statement is from http://pfam.wustl.edu/help-scores.shtml.

Zscore	No. of BCCs	No. of proteins	No. of APs
100	330	1758	295
110	731	1504	452
120	619	1317	237
130	530	1249	118
140	516	1208	99
150	514	1188	94
160	511	1171	83
170	496	1148	66
180	487	1111	63
190	483	1092	61
200	470	1076	42

Table 2. The number of biconnected components (BCCs), the number of sequences, and the number of articulation points (APs) at each Zscore interval of 10 from 100 to 200. The number of BCCs is the maximum at Zscore of 110. The higher number of BCCs is desirable but too many APs implies that many BCCs need to be merged into one.

Among 42 articulation points, 8 failed for AP-TEST, which implies families around these articulation points do share some domains in common. Among 470 families, 10 failed for RANGE-TEST, which implies that each subgraph has multiple families. We used $I_{split} = 50$ for splitting families and $I_{merge} = 10$ for merging families.

4.7.1 Splitting families

The list of six families failed for RANGE-TEST is shown in Table 3. These families are expected to have multiple domains, which will lead to the failure for RANGE-TEST. Three families (7, 18, and 58) are confirmed to have more than one domains detected by PFAM search (see Table 3) and the remaining three families (452, 454, and 465), which do not have domains

Family Name	Sequences	known domains
7	3322641 3322724 3322929 2687931 2688217 2688317 2688606	CheW CheR
18	3322451 3322737 3322802 3323075 3323208 2687964 2688415 2688449 2688636 2688747	GTP_EFTU GTP_EFTU_D2 GTP_EFTU_D3 EFG_C
58	3322341 3322643 3322811 3322930 2688314 2688460 2688488 2688604 2688707	Sigma54_activat response_reg HTH_8 CheB_methylest GGDEF
452	3322394 3322593 3322594 3322915 3322924 3322925	
454	3322399 3322400 3322413 3322755 3322756	
465	3322844 3323176 3323177 3323179 3323180 3323181	

Table 3. The list of six families failed for RANGE-TEST. All families are confirmed to have more than one domains detected by PFAM search or alignment in Figure 8. The sequence numbers are **gi** numbers from GENBANK. The domain names are from PFAM databases: CheW stands for CheW-like domain, CheR for CheR methyltransferase, GTP_EFTU for Elongation factor Tu GTP binding domain, GTP_EFTU_D2 for Elongation factor Tu domain 2, GTP_EFTU_D3 for Elongation factor Tu C-terminal domain, EFG_C for Elongation factor G C-terminus, Sigma54_activat for Sigma-54 interaction domain, response_reg Response regulator receiver domain, HTH_8 for Bacterial regulatory protein Fis family, CheB_methylest for CheB methylesterase, and GGDEF for GGDEF domain.

detected by PFAM search, are confirmed by sequence alignment as shown in Figure 8. The CLUSTER-SPLIT procedure split the six families into subfamilies of the same functional domains as shown in Table 4. After the splitting step, there were 480 families and they were merged with the HYPERGRAPH-MERGE procedure (explained in the following section).

Family	Zscore	Split families	Sequences	Common domains
7	250	7.1	2688217 2688606 3322724 **3322641**	CheW
		7.2	2687931 2688317 3322929 **3322641**	CheR
18	300	18.1	2687964 2688449 2688636 3322737 3322802 3323075	GTP_EFTU EFG_C GTP_EFTU_D2
		18.2	2688415 3322451	GTP_EFTU GTP_EFTU_D2
		18.3	2688747 3323208	GTP_EFTU GTP_EFTU_D2 GTP_EFTU_D3
58	250	58.1	2688707 3322341 **3322811**	Sigma54_activat
		58.2	**2688488 3322811**	response_reg
		58.3	**2688488 3322643**	response_reg
		58.4	2688460 2688604 **3322643**	response_reg
452	250	452.1	3322593 3322915 **3322925**	
		452.2	3322394 3322594 3322924 **3322925**	
454	450	454.1	3322756 **3322400**	
		454.2	3322399 3322413 3322755 **3322400**	
465	350	465.1	3323180 **3323181**	
		465.2	3323179 **3323181**	
		465.3	3322844 3323176 3323177 **3323181**	

Table 4. The families failed for RANGE-TEST were separated into subfamilies of the same functional domains (domains not common in the subfamilies are not shown). Note that splitting occurred at different Zscore values. Our iterative splitting procedure with stricter scores was highly effective. For example, splitting of Family 18 and 58 dealt with 4 different functional domains. Subfamilies, 58.2, 58.3, and 58.4, are merged while HYPERGRAPH-MERGE is performed. The sequence numbers in boldface denote multidomain proteins which belong to multiple families, i.e., articulation points in the sequence graph. Note that some multidomain proteins do not belong to multiple families. For example, all proteins in the Family 18 are multidomain proteins but belong to a single family.

4.7.2 Merging families

A hypergraph was formed as described in Section 4.5.6. There were 46 biconnected components, in each of which all families are considered for merging. To distinguish the biconnected components in the hypergraph from those in the sequence graph, we will denote the biconnected component in the hypergraph as *BCH*. Among 46 BCHs, 18 were further merged

Figure 8. Sequence alignments for the three families (452, 454, and 465) failed for RANGE-TEST but with no domains detected by PFAM. The alignment is with respect to the first sequence in the alignment which is a multidomain protein.

iteratively. In total, there were 64 cluster merging events that can be classified into four different types. Here, we describe each example of the four distinct types. The four examples for merging families are summarized in Table 5.

The first type of merging is to simply merge all families in a BCH. For example, all proteins in a BCH with Families 68 and 69, are merged into a new family.

The second type of merging is the same as the first type in terms of the merging procedure. However, this type of merging involves families that were previously split by the CLUSTER-SPLIT procedure. Family 7.1 (see Table 4) and Family 6 were merged into one and all proteins in the merged family share CheW domain.

The third type of merging includes only a part of families among those considered for merging. Family 134-135 was formed by merging two

New family	Round	Families	Zscore	Sequences	Common domains
68-69	1	68, 69	190	3322777 2688501 2688521 2688621 2688620 3322296 3322938 3322939 2688522	MCPsignal
6-7.1	1	6, 7.1	190	2688606 3322724 **3322641** 2688217 3322724 2688491	CheW
134 -135	1	133, 134, 135	150	3322552 3323244 2688008	S1
369-370-371	1	369, 370	190	3322962 3322518 3322963	
	2	369 - 370 371	170	3322962 3322518 3322963 2688608	

Table 5. Four types of merging events. The first type of merging, the new family 68-69, is simply to merge all families in a biconnected component of a hypergraph. The second type of merging, the new family 6-7.1, is the same merging procedure as the first type, but involves families that were split in the splitting step. The third type of merging, the new family 134-135, includes only part of families that were considered for merging. The forth type of merging, the new family 369-370-371, is from iterative merging processes of biconnected components, merging Families 369 and 370 and then merging with Family 371.

families, 134 and 135. However, the articulation point 3322552 belongs to Family 133 as well as 134 and 135. In Section 4.5.2, we discussed the type of merging in detail.

The fourth type of merging requires recursive merging of families. For example, two BCHs, one with Families 369 and 370 and the other with Families 370 and 371, were merged into a single family of proteins; merging of Families 369 and 370 and then merging with Family 371. All proteins in the resulting family are annotated as flagellar filament outer layer protein, but PFAM search did not find any domain.

Among 64 merging attempts, seven failed to be merged and all proteins in multiple families in the seven BCHs are shown to have multiple domains in Table 6. In total, 441 families with 1,076 sequences were classified.

Sequence	Families	Multiple domains or ranges
2688314	57,58	response_reg, GGDEF
3322930	58,59	response_reg, CheB_methylest
3322379	64,65	helicase_C, UVR
2688004	138,139	(27,625)(744,895)
3322920	151,152	(42,316)(341,474)(OmpA)
3322260	445,447	(1,125)(140,206)
3322272	446,447	(27,93)(114,194)

Table 6. The seven multidomain proteins detected by failure in merging families. The numbers in parentheses denote the ranges that are shared among proteins in the family. Module names are from PFAM search: response_reg for Response regulator receiver domain, GGDEF for GGDEF domain, CheB_methylest for CheB methylesterase, helicase_C for Helicase conserved C-terminal domain, UVR for UvrB/uvrC motif, and OmpA for OmpA family.

4.8 Conclusion

As more sequences become available at an exponential rate, sequence analysis on a large number of sequences will be increasingly important. Sequence clustering algorithms are computational tools for that purpose. In this chapter, we surveyed the recent developments in clustering algorithms based on graph theory and presented our clustering algorithm, BAG, which used two graph properties, biconnected components and articulation points. Among the graph structures used in all five algorithms, the structure used in our algorithm, biconnected component, is weaker than those used in other algorithms. For example, the structure of triangular relationships used in COG and GeneRAGE is biconnected but not *vice versa*. Matsuda *et al.* [1999] and Matsuda [2000] compute globally optimal structures, *p*-quasi completeness and maximal density, respectively. In contrast, our algorithm is greedy and computes a local optimum. However, our algorithm utilizes the computational efficiency, i.e., linear time complexity, to achieve clustering of families of very specific categories. In particular, our algorithm was successful in classifying families where the relationships among member sequences were defined at different scores; for example, Families 7.1 and 7.2

can be separated at Zscore 250 but not at Zscore 200 where most of families were classified (see Table 4).

The future work for our algorithm includes applications of our algorithm to different types of sequences such as DNA and EST sequences. It would also be interesting to retain the hierarchical structure of the merging procedure so that sequence relationships can be seen at different levels. In addition, refining further each family in the context of genome, i.e., orthologs as used in COG, is an interesting topic for further research.

Acknowledgments

This work evolved from a project of the author at DuPont Central and Development. The basic algorithm in Section 4.4.1 was originally implemented with PERL graph package [Orwant, 1999] while working at DuPont. In addition, this work benefitted from many productive discussions with my colleagues in the microbial genomics project at DuPont. I thank my colleagues at DuPont, especially Li Liao and Jean-Francois Tomb. Li Liao introduced to me the two graph theoretic clustering algorithms [Matsuda *et al.*, 1999] and [Matsuda, 2000].

The work presented in this chapter was conducted at Indiana University and supported by the INGEN (Indiana Genomics) initiative. The pairwise comparisons of proteins from 63 completed genomes were performed using the IBM SP supercomputer at Indiana. The Indiana University Computing Center graciously supported this project. I especially thank Craig Stewart and Mary Papakhian for their support.

I thank Narayanan Perumal, Mehmet Dalkilic, and Dennis Groth of Indiana University School of Informatics for their reading this chapter and their suggestions.

References

Altschul, S.F., Gish, W., Miller, W., Myers, E.W. and Lipman, D.J. (1990) "Basic local alignment search tool." *Journal of Molecular biology* **215,** 403-410.

Altschul, S.F., Madden, T.L., Schäffer, A.A., Zhang, J., Zhang, Z., Miller, W. and Lipman, D.J. (1997) "Gapped BLAST and PSI-BLAST: a new generation of protein database search programs." *Nucleic Acids Research* **25,** 3389-3402.

Bateman, A., Birney, E., Durbin, R., Eddy, S. R., Howe, K. L. and Sonnhammer, E. L. L. (2000) "The Pfam Protein Families Database." *Nucleic Acids Research* **28**, 263-266.

Cannarozzi, G., Hallett, M.T., Norberg, J. and Zhou, X. (2000) "A cross-comparison of a large dataset of genes." *Bioinformatics* **16**, 654-655.

Cormen, T. H., Leiserson, C. E. and Rivest, R. L. (1989) *Introduction to Algorithms.* MIT Press.

Enright, A. J. and Ouzounis, C. A. (2000) "GeneRAGE: A robust algorithm for sequence clustering and domain detection." *Bioinformatics* **16**, 451-457.

Gilbert, D. G. (2002) "euGenes: A eukaryote genome information system." *Nucleic Acids Research* **30**, 145-148.

Matsuda, H., Ishihara, T. and Hashimoto, A. (1999) "Classifying molecular sequences using a linkage graph with their pairwise similarities." *Theoretical Computer Science* **210**, 305-325.

Matsuda, H. (2000) "Detection of conserved domains in protein sequences using a maximum-density subgraph algorithm." *IEICE Transactions Fundermentals* **E83-A**, 713-721.

Mehlhorn, K. and Naher, S. (1999) *LEDA: A Platform for Combinatorial and Geometric Computing.* University of Cambridge Press.

Orwant, J., Hictaniemi, J. and Macdonald, J. (1999) *Mastering Algorithms with Perl.* O'reilly.

Park, J., Teichmann, S.A., Hubbard, T. and Chothia, C. (1997) "Intermediate sequences increase the detection of homology between sequences." *Journal of Molecular Biology* **273**, 349-354.

Pearson, W. R. and Lipman, D. J. (1988) "Improved tools for biological sequence comparison." *Proc. National Academy of Science* **85**, 2444-2448.

Siek, J. G., Lee, L.Q. and Lumsdaine, A. (2002) *Boost Graph Library: The User Guide and Reference Manual.* Addison-Wesley. www.boost.org.

Smith, T. F. and Waterman, M. F. (1981) "Identification of common molecular subsequences." *Journal of Molecular Biology* **147**, 195-197.

Tatusov, R. L., Koonin, E. V. and Lipman, D. J. (1997) "A genomic perspective on protein families." *Science* **278**, 631-637.

Tatusov, R. L., Natale, D. A., Garkavtsev, I. V., Tatusova, T. A., Shankavaram, U.
T., Rao, B. S., Kiryutin, B., Galperin, M. Y., Fedorova, N. D. and Koonin, E. V.
(2001) "The COG database: New developments in phylogenetic classification of
proteins from complete genomes." *Nucleic Acids Research* **29,** 22-28.

Author's Address

Sum Kim, School of Informatics and Center for Genomics and
Bioinformatics, Indiana University, Bloomington, USA.
Email: sunkim@bio.informatics.indiana.edu, sun.kim@acm.org.

Chapter 5

The Protein Information Resource for Functional Genomics and Proteomics

Cathy H. Wu

5.1 Introduction

The human genome project has revolutionized the practice of biology and the future potential of medicine. The draft DNA sequence of the human genome has been published [McPherson *et al.*, 2001; Venter *et al.*, 2001], and complete genomes of other organisms continue to be sequenced *en masse*. Meanwhile, there is growing recognition that proteomic studies bring the researcher closer to the actual biology than studies of gene sequence or gene expression alone. High-throughput studies are being conducted and rapid advances being made in areas such as protein expression, protein structure and function, and protein-protein interactions. Given the enormous increase in genomic, proteomic, and molecular data, computational approaches, in combination with empirical methods, are expected to become essential for deriving and evaluating hypotheses. To fully explore these valuable data, advanced bioinformatics infrastructures must be developed for biological knowledge management. One major challenge lies in the volume, complexity, and dynamic nature of the data, which are being collected and maintained in heterogeneous and distributed sources. New approaches need to be devised for data collection, maintenance, dissemination, query, and analysis. The Protein Information Resource (PIR) [Wu *et al.*, 2002] aims to serve as an integrated public resource of functional annotation of proteins to support genomic/proteomic research and scientific discovery. It provides many protein databases and data analysis tools, and employs family classification approach to facilitate exploration of proteins and comparative studies of various family relationships. Such knowledge is fundamental for our understanding of protein evolution, structure, and function.

The PIR was established in 1984 as a resource to assist researchers in the identification and interpretation of protein sequence information. The PIR, along with the Munich Information Center for Protein Sequences (MIPS) [Mewes *et al.*, 2000], and the Japan International Protein Information Database (JIPID), continues to enhance and distribute the PIR-International Protein Sequence Database. The database evolved from the first comprehensive collection of macromolecular sequences in the *Atlas of Protein Sequence and Structure* published under the editorship of Margaret O. Dayhoff [1965], who pioneered molecular evolution research.

Central to the organization and annotation of the PIR databases are protein family and domain relationships. Protein family classification is well recognized as an effective approach for large-scale analysis of genomic sequences and for functional annotation of proteins. We also utilize the classification approach for database organization and integration of protein sequence, structure, and function. Major protein family organizations include hierarchical families of proteins, such as PIR superfamilies [Barker *et al.*, 1996]; families of protein domains, such as those in Pfam [Bateman *et al.*, 2000]; sequence motifs or conserved regions, as in ProSite [Hofmann *et al.*, 1999]; and integrated family classification, as in iProClass [Wu *et al.*, 2001].

To further support functional genomic and proteomic research, we have greatly improved our bioinformatics infrastructure in the last three years, which allows us (i) to continue to provide high quality protein sequence data and annotation, while keeping pace with the large influx of data being generated by genome sequencing projects, (ii) to develop an integrated system of protein databases and analytical tools for expert annotation and knowledge discovery, and (iii) to improve accessibility of our resource and interoperability of our databases. Some key developments include: highly automated protein sequence classification and annotation, new submission mechanism for bibliography data, new non-redundant reference protein database to provide timely sequence collection, new integrated classification database to provide comprehensive protein information, database migration into Oracle 8i object-relational database system, database distribution in XML format, and redesign of the web site for easy navigation, information retrieval, and sequence analysis.

5.2 PIR-International Protein Sequence Database

The PIR-International Protein Sequence Database (PSD) is a highly annotated and classified protein sequence database in the public domain. It currently (March 2002) contains more than 283,000 protein sequences with comprehensive coverage across the entire taxonomic range, including sequences from publicly available complete genomes.

Superfamily Classification

A unique characteristic of the PIR-PSD is the superfamily classification that provides complete and non-overlapping clustering of proteins based on global (end-to-end) sequence similarity. Sequences in the same superfamily share common domain architecture (i.e., have the same number, order, and types of domains) and do not differ excessively in overall length unless they are fragments or result from alternate splicing or initiators. The automated classification system places new members into existing superfamilies and defines new superfamily clusters using parameters including the percentage of sequence identity, overlap length ratio, distance to neighboring superfamily clusters, and overall domain arrangement. Currently, over 99% of sequences are classified into families of closely related sequences (at least 45% identical), and over two thirds of sequences are classified into >36,000 superfamilies. The automated classification is being augmented by manual curation of superfamilies, starting with those containing at least one definable domain, to provide superfamily names, brief descriptions, bibliography, list of representative and seed members, as well as domain and motif architecture characteristic of the superfamily. Sequences in PIR-PSD are also classified with homology domains and sequence motifs. Homology domains, which are shared by more than one superfamily, may constitute evolutionary building blocks, while sequence motifs represent functional sites or conserved regions.

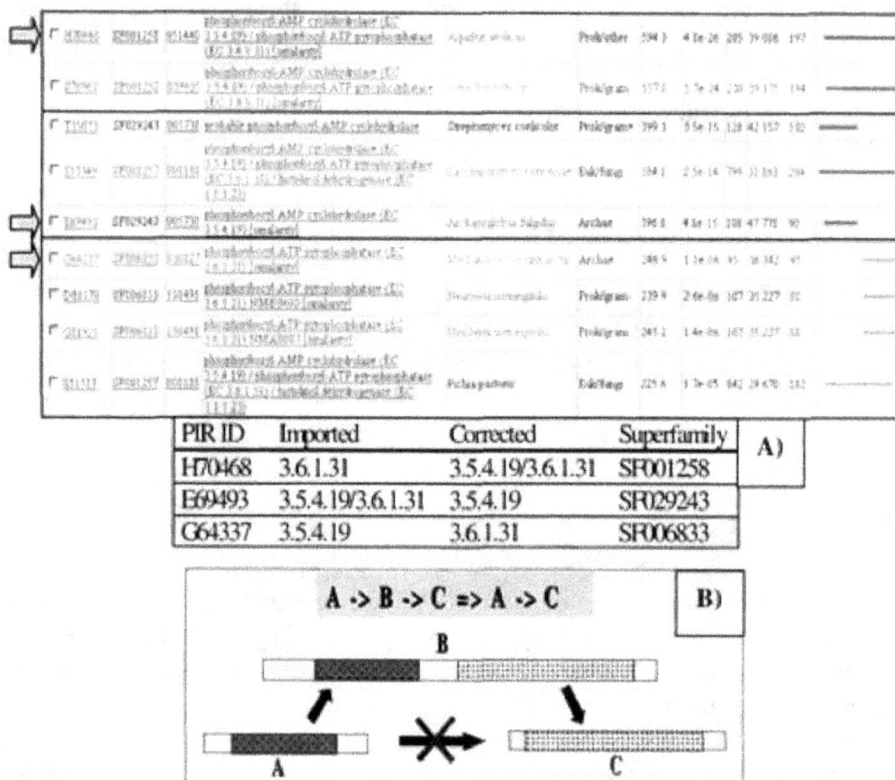

Figure 1. Genome sequence annotation - transitive catastrophe: (A) mis-annotation of imported entries corrected based on superfamily classification; (B) transitive identification error involving multi-domain proteins.

The classification allows systematic detection of genome annotation errors based on comprehensive superfamily and domain classification. Several annotation errors originated from different genome centers have lead to the so-called "transitive catastrophe." Figure 1 illustrates an example where several members of three related superfamilies were originally mis-annotated, likely because only local domain relationships were considered. Here, the related superfamilies are: SF001258 (hisI-bifunctional enzyme), a bifunctional protein with two domains for EC 3.5.4.19 and 3.6.1.31; SF029243 (phosphoribosyl-AMP cyclohydrolase), containing only the first domain for EC 3.5.4.19; and SF006833 (phosphoribosyl-ATP pyrophosphatase), containing the second domain for EC 3.6.1.31. Based on the superfamily classification, the improper names assigned to three sequence entries imported to PIR (H70468, E69493, G64337) were later corrected (Figure 1A). The type of transitive identification error observed in entry G64337 (named as EC 3.5.4.19 when it is actually EC 3.6.1.31) often involves multi-domain proteins (Figure 1B).

The classification also provides the basis for rule-based procedures that are used to propagate information-rich annotations among similar sequences and to perform integrity checks. These scripts use the superfamily/family classification system and sequence patterns and profiles to produce highly specific annotations. False positives are avoided by applying automated annotations only to classified members of the families and superfamilies for which the annotation has been validated. Integrity checks are based on PIR controlled vocabulary, standard nomenclature (such as IUBMB Enzyme Nomenclature, http://www.chem.qmw.ac.uk/iubmb/enzyme/), and the saurus of synonyms or alternate names.

Evidence Attribution and Bibliography Submission

Attribution of protein annotations to validated experimental sources provides effective means to avoid propagation of errors that may have resulted from large-scale genome annotation. To distinguish experimentally verified from computationally predicted data, PIR-PSD entries are labeled with status tags of *"validated"*, *"similarity"*, or *"imported"* in protein Title, Function and Complex annotations (Figure 2A). The entries are also tagged with *"experimental"*, *"predicted"*, *"absent"*, or *"atypical"* in Feature annotations (Figure 2B). The *validated* Function or Complex annotation includes hypertext-linked PubMed unique identifiers for the articles in which the experimental determinations are reported.

Linking protein data to more literature data that describes or characterizes the proteins is crucial for increasing the amount of experimental information and improving the quality of protein annotation. We have developed a bibliography submission system for the scientific community to submit, categorize, and retrieve literature information for PSD protein entries. The submission interface guides users through steps in mapping the paper citation to given protein entries and entering the literature data. The submission form includes a section where the literature data are summarized using categories

ENTRY	T48678	**(A)**
TITLE	proteasome alpha-1 chain [**validated**] - Haloferax volcanii	
COMPLEX	heterodimer; alpha-1 and beta-1 (PIR:T48677) chain [**validated; PMID:10482525**]	
FUNCTION	#description the predominant peptide-hydrolyzing activity of the alpha (1)beta(1)-proteasome is cleavage carboxyl to hydrophobic residues [**validated; PMID:10482525**]	

ENTRY	XNHUSP #type complete	**(B)**
TITLE	serine--pyruvate transaminase (EC 2.6.1.51), peroxisomal - human	
FEATURE		
2-392	#product serine--pyruvate transaminase, peroxisomal #status **experimental** #label MAT\	
390-392	#region peroxisome/glyoxysome location signal #status **atypical**\	
2	#modified_site acetylated amino end (Ala) (in mature form) #status **experimental**\	
209	#binding_site pyridoxal phosphate (Lys) (covalent) #status **predicted**\	
367	#binding_site carbohydrate (Asn) (covalent) #status **absent**	

Figure 2. Evidence attribution tags in PIR for (A) Title, Complex, and Function annotation, and (B) Feature annotation.

(such as genetics, tissue/cellular localization, molecular complex or interaction, function, regulation, and disease), with evidence attribution (experimental or predicted) and description of methods. Also included is a literature information page that provides data mining and displays both references cited in PIR and submitted by users.

5.3 PIR Non-Redundant Refence Protein Database

The PIR-NREF (Non-redundant REFerence) protein database is designed to provide a timely and comprehensive collection of all protein sequence data, keeping pace with the genome sequencing projects and containing source attribution and minimal redundancy. The database has three major features: (i) *comprehensiveness and timeliness*: it currently consists of about 900,000 sequences from PIR-PSD, SwissProt [Bairoch and Apweiler, 2000], TrEMBL, RefSeq [Pruitt and Maglott, 2001], GenPept, and PDB [Berman *et al.*, 2000], and is updated biweekly; (ii) *non-redundancy*: it is clustered by sequence identity and taxonomy at the species level; and (iii) *source attribution*: it contains protein IDs and names from associated databases in addition to protein sequence, taxonomy, and bibliography.

As illustrated in Figure 3, each NREF entry represents an identical amino acid sequence from the same source organism redundantly presented in one or more underlying protein databases. The NCBI taxonomy (http://www.ncbi.nlm.nih.gov/Taxonomy/taxonomyhome.html/) is used as the ontology for matching source organism names at the species or strain (if known) levels. The *Thioredoxin* sample entry report (Figure 3) shows that identical sequences (of 100% sequence identity and identical length) are found in two strains, K12 and B, of *Escherichia coli*. The report also displays identical sequences from different species or sources in the "*Related Sequence*" section, including one from *Salmonella typhimurium*. The section will soon present closely related NREF sequences and identical substrings.

The NREF database can be used to assist functional identification of proteins, ontology development of protein names, and detection of annotation errors or discrepancies. Comprehensive, non-redundant, and with source attribution, NREF is an ideal underlying database for sequence analysis tasks. The clustering with source organisms supports analysis of evolutionary relationships of proteins and allows easy compilation of sub-databases based on taxonomy to refine sequence searches. The composite protein names from all underlying protein databases, including synonyms, alternate names, and even misspellings, constitute an initial thesaurus of terms that can help ontology development of protein names. For example (Figure 4A), a protein may be variably named based on function at different hierarchical levels (ATP-dependent RNA helicase vs. RNA helicase), gene name (protein p68), motif sequence similarity (DEAD/H box-5), combinations of function and gene name (RNA helicase p68), and other combinations. The NREF database also provides composite bibliography information with PubMed cross-references for direct online abstract retrieval. Together, the database and the abstracts provide an important knowledge base for applying computational linguistics or natural language processing technologies to the problem of

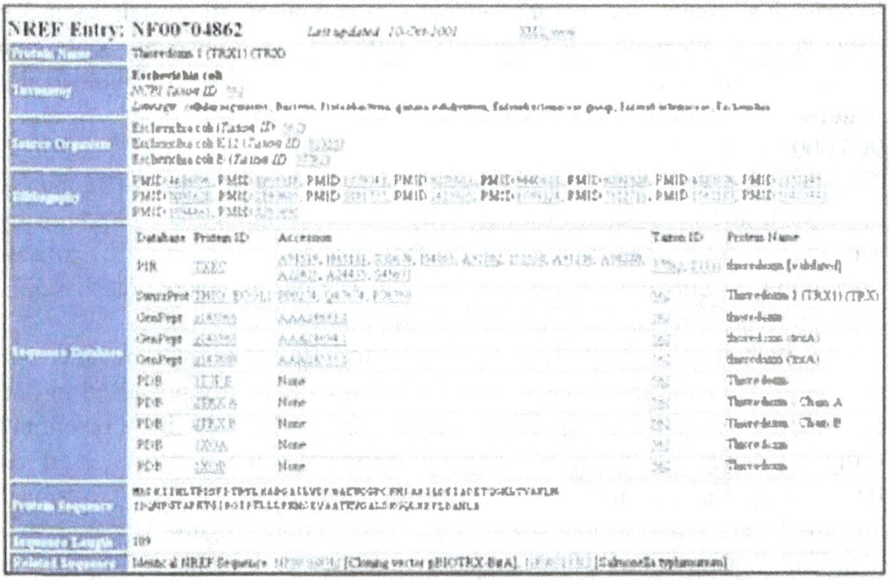

Figure 3. PIR-NREF entry report with attribution of source protein databases.

Figure 4. PIR-NREF composite protein names for (A) ontology development of protein names, and (B) detection of discrepant and/or incorrect annotations.

protein name ontology [Yoshida *et al.*, 2000]. The different protein names assigned by different databases may also reflect annotation discrepancies. As an example (Figure 4B), the protein (PIR: T40073) is variously named as a monofunctional (EC 3.5.4.19), bifuntional (EC 3.5.4.19, 3.6.1.31), and trifunctional (EC 3.5.4.19, 3.6.1.31, 1.1.1.23) protein. The source name attribution, thus, provides clues for potentially mis-annotated proteins.

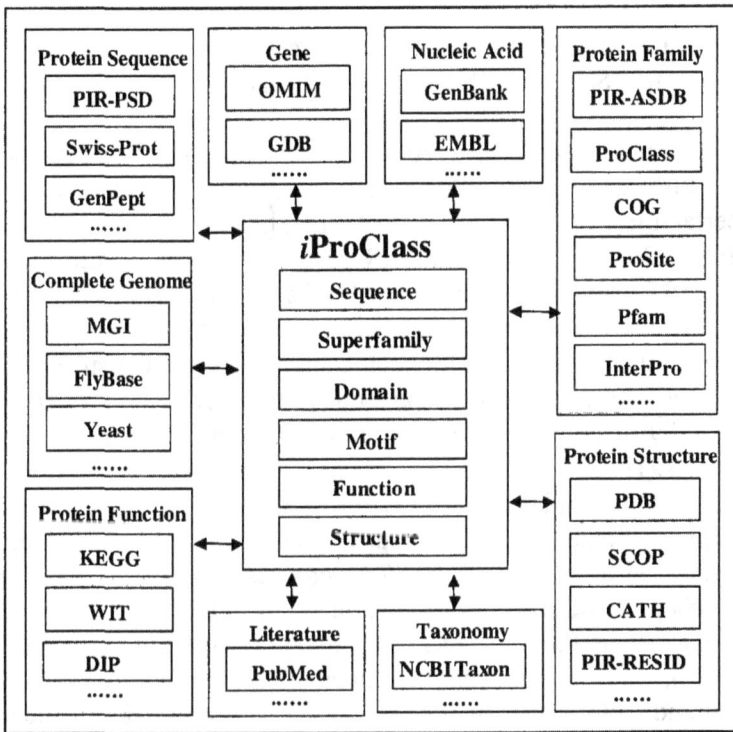

Figure 5. iProClass database for data integration: modular architecture and extensive links.

5.4 Integrated Protein Classification Database

The iProClass (integrated Protein Classification) database (Figure 5) is designed to provide comprehensive descriptions of all proteins and serve as a framework for data integration in a distributed networking environment. It is extended from ProClass [Wu *et al.*, 1996; Huang *et al.*, 2000], a protein family database that organizes proteins based on PIR superfamilies and ProSite motifs. The protein information in iProClass includes family relationships at both global (/family) and local (domain, motif, site) levels, as well as structural and functional classifications and features of proteins. A modular architecture organizes the information into multiple database components for Sequence, Superfamily, Domain, Motif, Structure, and Function.

The current version (March 2002) consists of more than 320,000 non-redundant PIR-PSD and SwissProt proteins organized with more than 36,000 PIR superfamilies, 100,000 families, 3700 PIR homology and Pfam domains, 1300 ProSite/ProClass motifs, 280 PIR post-translational modification sites, 250,000 FASTA similarity clusters, and links to over 45 molecular biology databases. The post-translational modifications are documented in the RESID database [Garavelli *et al.*, 2001], which contains information such as names, formula, molecular weights, and links to PSD entries containing experimentally determined or computationally predicted modifications with evidence tags. The FASTA similarity clusters are collected in the PIR-ASDB (Annotation and Similarity DataBase) [McGarvey *et al.*, 2000], which contains pre-computed, biweekly-updated sequence neighbors of all PSD entries based on all-against-all FASTA searches [Pearson and Lipman, 1988]. Other iProClass cross-references include databases for protein families (e.g., Pfam, ProSite, COG [Tatusov *et al.*, 2001], InterPro [Apweiler *et al.*, 2001]), enzymes, functions, and interactions (e.g., EC-IUBMB, KEGG [Kanehisa and Goto, 2000], WIT [Overbeek *et al.*, 2000], DIP [Xenarios *et al.*, 2001]), structures and structural classifications (e.g., PDB, SCOP [Lo Conte *et al.*, 2000], CATH [Pearl *et al.*, 2001], PDBSum [Laskowski, 2001]), genes and genomes (e.g., TIGR [Peterson *et al.*, 2001], OMIM [Wheeler *et al.*, 2001]), ontologies (e.g., Gene Ontology [Ashburner *et al.*,

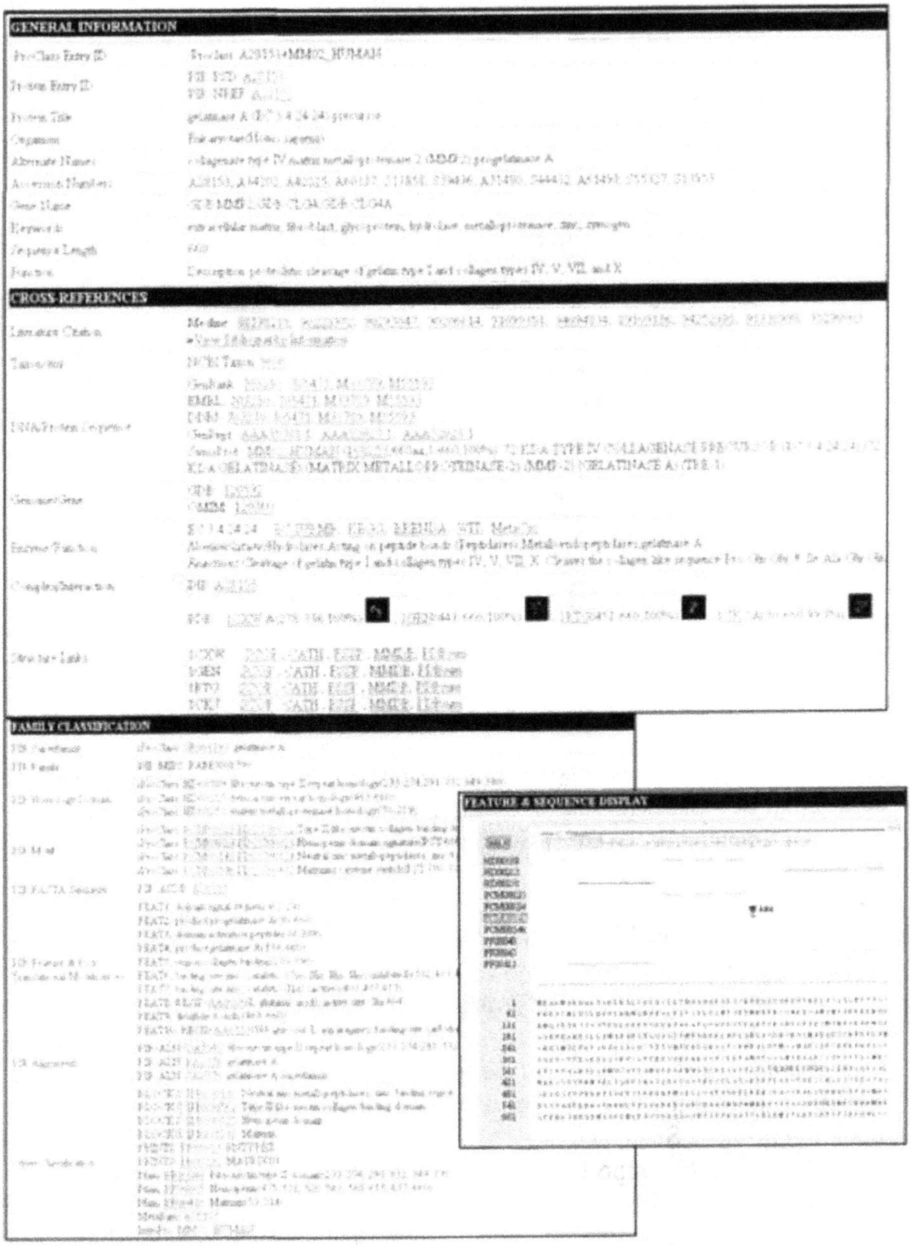

Figure 6. iProClass protein sequence entry report, example retrievable at http://pir.georgetown.edu/cgi-bin/iproclass/iproclass?choice=entry&id=A28153.

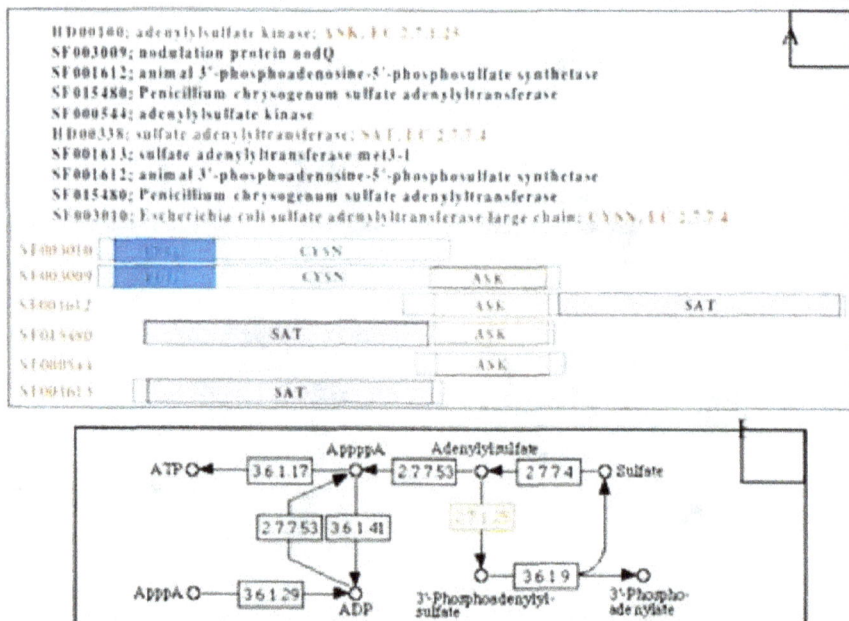

Figure 7. Superfamily-domain-function relationship to reveal protein functional association: (A) association of ASK (EC 2.7.1.25) and SAT/CYSN (EC2.7.7.4) in multi-domain proteins; (B) their association in a metabolic pathway.

2000]), literature (NCBI PubMed, http://www.ncbi.nlm.nih.gov/Literature/), and taxonomy (NCBI Taxonomy).

The extensive protein information is organized in Sequence report (Figure 6) in four sections, *General Information, Cross-References, Family Classification*, and *Feature and Sequence Display*, with hypertext links for further exploration and graphical display of domain and motif regions. Built upon the primary Sequence reports are views of protein family relationships. The Superfamily report provides summaries including membership information with length, taxonomy, and keyword statistics, complete member listing separated by major kingdoms, family relationships, and structure and function cross-references.

To be further implemented are the Domain-Motif components that represent domain and motif-centric views with direct mapping to superfamilies, the Function component that describes functional properties of enzymes and other activities, and relationships such as families, pathways, and processes, as well as the Structure component that describes structural properties and relates structural classes to evolution and function. Such data integration is important in revealing protein functional associations beyond sequence homology, as illustrated in the following example. As shown in Figure 7A, the ASK domain (EC 2.7.1.25) appears in four different superfamilies, all having different overall domain arrangements. Except for SF000544, the other three superfamilies are bifunctional, all containing sulfate adenylyltransferase (SAT) (EC 2.7.7.4). However, the same SAT enzymatic activity is found in two distinct sequence types, the SAT domain and CYSN homology. Furthermore, both EC 2.7.1.25 and EC 2.7.7.4 are in adjacent steps of the same metabolic pathway (Figure 7B). This example demonstrates that protein function may be revealed based on domain and/or pathway association, even without obvious sequence homology. The iProClass database design would present such complex superfamily-domain-function relationship to assist functional identification or characterization of proteins.

A key design objective of the iProClass database system is to address the database interoperability issue arising from the voluminous, heterogeneous, and distributed data [Davidson *et al.*, 1995]. There are several general approaches for data integration. The iProClass uses database links as a foundation for interoperability [Karp, 1995] and combines both data warehouse and hypertext navigation methods. In our approach, we restrict the database content to the immediate needs of protein analysis and annotation and store a rich collection of links with related summary information. The latter will alleviate potential problems associated with timely collection of information from distributed sources over the Internet. The idea is similar to that of the Virgil database [Achard *et al.*, 1998], which was developed to model the concept of rich links (the link itself and the related pieces of information) between database objects. Another iProClass design principle that promotes database interoperation is the adoption of a modular and open architecture. The modular structure makes the system scalable, customizable and extendable for adding new components. The open framework with common database schema, data format, and query interface allows data sharing among distributed research groups and integration of other database components important for genomic and proteomic studies. For

example, the iProClass Protein Sequence module can be linked to Gene
Expression and Protein Expression modules via gene-protein mapping and
peptide-protein mapping, respectively.

5.5 PIR System Distribution

PIR Web Access

The PIR web site (http://pir.georgetown.edu) [McGarvey *et al.*, 2000]
connects data mining and sequence analysis tools to underlying databases for
protein information retrieval and knowledge discovery. The site has been
redesigned to include a user-friendly navigation system and more graphical
interfaces and analysis tools. The Major PIR pages are listed in Table 1.

The PIR-PSD interface provides entry retrieval, batch retrieval, basic or
advanced text searches, and various sequence searches. The PIR-NREF
interface supports direct report retrieval as well as full-scale sequence search
for list retrieval. Report retrieval is based on sequence unique identifiers,

Description	Web Page URL
PIR Home	http://pir.georgetown.edu
PIR-PSD	http://pir.georgetown.edu/pirwww/search/textpsd.shtml
iProClass	http://pir.georgetown.edu/iproclass
PIR-NREF	http://pir.georgetown.edu/pirwww/search/pirnref.shtml
PIR-ASDB	http://pir.georgetown.edu/cgi-bin/asdblist.pl?id=H70468
Bibliography	http://pir.georgetown.edu/pirwww/literature.html
PIR databases	http://pir.georgetown.edu/pirwww/dbinfo/dbinfo.html
PIR searches	http://pir.georgetown.edu/pirwww/search/searchseq.html
FTP site	ftp://nbrfa.georgetown.edu/pir_databases/

Table 1. Major PIR web pages for data mining and sequence analysis.

including the NREF ID and sequence unique identifiers of the source databases. Several sequence search options are available for functional identification of proteins and peptides, including BLAST [Altschul *et al.*, 1997] Search, Peptide Match, and Pattern Match. The BLAST Search of a user-supplied query sequence against NREF sequences returns a list of all matched sequences above a given threshold. As shown in the example (http://pir.georgetown.edu/iproclass/NFBLASTex.html), for each matched database sequence, information is provided for NREF ID, protein IDs and names from associated databases (with hypertext links for retrieval of up-to-date source entries), organism name, and sequence match result with scores and visualization. The Peptide Match finds an exact match in the NREF database to a user-defined peptide sequence. The Pattern Match searches for a user-defined pattern or ProSite pattern against all NREF sequences. The iProClass interface includes both sequence and text searches. The BLAST Search returns best-matched proteins and superfamilies, each displayed with a one-line summary linking to complete reports. Peptide Match allows protein identification based on peptide sequences. Text Search supports direct search of the underlying Oracle tables using unique identifiers or combinations of text strings, based on a Java program running JDBC. The FASTA clusters are directly retrievable from the web interface based on PIR ID (e.g. http://pir.georgetown.edu/cgi-bin/asdblist.pl?id=H70468), where neighbors are listed with annotation information and graphical displays of sequence similarity matches.

Other sequence searches supported on the PIR web site include hidden Markov model [Eddy *et al.*, 1995] search for PIR homology and Pfam domains and ProSite motifs, Smith-Waterman [1981] pair-wise alignment, ClustalW [Thompson *et al.*, 1994] multiple alignment, IESA (Integrated Environment for Sequence Analysis) [McGarvey *et al.*, 2000], and GeneFIND [Wu *et al.*, 1999] family identification.

PIR FTP Transfer

The PIR anonymous FTP site (ftp://nbrfa.georgetown.edu/pir_databases) provides direct file transfer. Files distributed include the PIR-PSD, PIR-NREF, other auxiliary databases, other documents, files, and software programs. The PIR-PSD has been distributed as flat files in NBRF and CODATA formats. Both PSD and NREF data files are also distributed in

```
 - <NrefEntry id="NF00000001" update_date="27-Sep-2001">              A
     <protein_name>LECTIN (FRAGMENT)</protein_name>
 - <taxonomy>
     <species_name>Caragana arborescens</species_name>
     <taxon_id>20484</taxon_id>
     <lineage>cellular
        organisms;Eukaryota;Viridiplantae;Streptophyta;Charophyta/Embryophyta
        group;Embryophyta;Tracheophyta;Euphyllophyta;Spermatophyta;Magnoliophyta;
        eudicotyledons;core eudicots;Rosidae;eurosids
        I;Fabales;Fabaceae;Papilionoideae; Galegeae;Caragana</lineage>
   </taxonomy>
 - <source_organism>
    - <source_org>
       <organism_name>Caragana arborescens</organism_name>
       <taxon_id>20484</taxon_id>
     </source_org>
   </source_organism>
 - <seq_database>
    - <source_db db="GenPept">
       <protein_id>g3819121</protein_id>
       <protein_name>lectin</protein_name>
       <taxon_id>20484</taxon_id>
       <accession>CAA13596.1</accession>
     </source_db>
    + <source_db db="TrEMBL">
     </seq_database>
 - <protein_seq>
     <length>90</length>

     <sequence>VAVEFDTFCNRDWDPEHRHIGIDVNHISSVGTTAWNLSNGDVAAVEIIYHAVTHE
        GYDRSSRPIYVLKEKVDLRRYLPEWVRIGF</sequence>
   </protein_seq>
 </NrefEntry>
```

```
<!-- Entry: the root element. -->                                     B
<!ELEMENT NrefEntry    (protein_name, taxonomy?, source_organism,
                        bibliography?,seq_database,protein_seq, related_seq? )>
<!ATTLIST NrefEntry id            ID      #REQUIRED
                    update_date   CDATA #IMPLIED >


<!-- protein_name: The protein name. -->
<!ELEMENT protein_name   (#PCDATA) >        <!-- protein name -->

<!-- taxonomy: identification of the biological source. -->
<!ELEMENT taxonomy ( species_name, common_name?, taxon_id, lineage) >

<!ELEMENT species_name ( #PCDATA ) >      <!-- scientific species name -->
<!ELEMENT common_name  ( #PCDATA ) >      <!-- common name -->
<!ELEMENT taxon_id     ( #PCDATA ) >      <!-- NCBI taxonomy ID -->
<!ELEMENT lineage      ( #PCDATA ) >      <!-- taxonomy lineage -->

<!-- source_organism: identification of the source species,subspecies,strains or varietas. -->
<!ELEMENT source_organism  (source_org+)>

<!ELEMENT source_org       (organism_name,taxon_id)>

<!ELEMENT organism_name (#PCDATA)>        <!-- organism name -->


<!-- bibliography: related bibliography -->
<!ELEMENT bibliography ( pmid?, muid? )>

<!ELEMENT pmid  ( #PCDATA )   >  <!-- PubMed identifiers            -->
<!ELEMENT muid  ( #PCDATA )   >  <!-- MEDLINE unique identifiers-->


<!-- seq_database: sequence source database -->
<!ELEMENT seq_database   (source_db+)>

<!ELEMENT source_db     (protein_id,protein_name,taxon_id,accession)>
<!ATTLIST source_db     db  (PIR | Swissprot | TrEMBL | GenPept |
                         RefSeq | PDB)   #REQUIRED>
```

Figure 8. PIR-NREF entry (A) in XML format, (B) with an associated DTD (Document Type Definition) (partially shown).

XML format (Figure 8A) with associated DTD (Document Type Definition) (Figure 8B) files. The PSD and NREF sequences are available in FASTA format.

The PIR-PSD, iProClass, and PIR-NREF databases have been implemented in Oracle 8i object-relational database system on our Unix server. To enable open source distribution, the databases are being mapped to MySQL and ported to Linux system. Since February 2002, the PSD database has been distributed in MySQL. To establish reciprocal links to PIR databases, to host a PIR mirror web site, or to request PIR database schema, please contact pirmail@nbrf.georgetown.edu.

5.6 Conclusion

The PIR serves as a primary resource for exploration of proteins, allowing users to answer complex biological questions that may typically involve querying multiple sources. In particular, interesting relationships between database objects, such as relationships among protein sequences, families, structures, and functions, can be revealed readily. Functional annotation of proteins requires association of proteins based on properties beyond sequence homology - proteins sharing common domains connected via related multi-domain proteins (grouped by superfamilies); proteins in the same pathways, networks, or complexes; proteins correlated in their expression patterns; and proteins correlated in their phylogenetic profiles (with similar evolutionary patterns) [Marcotte *et al.*, 1999]. The PIR, with its integrated databases and analysis tools, thus constitutes a fundamental bioinformatics resource for biologists who contemplate using bioinformatics as an integral approach to their genomic/proteomic research and scientific inquiries.

Acknowledgments

PIR is a registered mark of National Biomedical Research Foundation. The PIR is supported by grant P41 LM05978 from the National Library of Medicine, National Institutes of Health. The iProClass and RESID databases are supported by DBI-9974855 and DBI-9808414 from the National Science Foundation.

References

Achard, F., Cussat-Blanc, C., Viara, E. and Barillot, E. (1998) "The new Virgil database: a service of rich links." *Bioinformatics* **14**, 342-348.

Altschul, S.F., Madden, T.L., Schaffer, A.A., Zhang, J., Zhang, Z., Miller, W. and Lipman, D.J. (1997) "Gapped BLAST and PSI-BLAST: A new generation of protein database search programs." *Nucleic Acids Research* **25**, 3389-3402.

Apweiler, R., Attwood, T. K., Bairoch, A., Bateman, A., Birney, E., Biswas, M., Bucher, P., Cerutti, L., Corpet, F., Croning, M. D., Durbin, R., Falquet, L., Fleischmann, W., Gouzy, J., Hermjakob, H., Hulo, N., Jonassen, I., Kahn, D., Kanapin, A., Karavidopoulou, Y., Lopez, R., Marx, B., Mulder, N. J., Oinn, T. M., Pagni, M., Servant, F. and Zdobnov, E. M. (2001) "The InterPro database, an integrated documentation resource for protein families, domains and functional sites." *Nucleic Acids Research* **29**, 37-40.

Ashburner, M., Ball, C. A., Blake, J. A., Botstein, D., Butler, H., Cherry, J. M., Davis, A. P., Dolinski, K., Dwight, S. S., Eppig, J. T., Harris, M. A., Hill, D. P., Issel-Tarver, L., Kasarskis, A., Lewis, S., Matese, J. C., Richardson, J. E., Ringwald, M., Rubin, G. M. and Sherlock, G. The Gene Ontology Consortium (2000) "Gene ontology: Tool for the unification of biology." *Nature Genetics* **25**, 25-29.

Bairoch, A. and Apweiler, R. (2000) "The Swiss-Prot protein sequence database and its supplement TrEMBL in 2000." *Nucleic Acids Research* **28**, 45-48.

Barker, W.C., Pfeiffer, F. and George, D. (1996) "Superfamily classification in PIR-international protein sequence database." *Methods in Enzymology* **266**, 59-71.

Bateman, A., Birney, E., Durbin, R., Eddy, S. R., Howe, K. L. and Sonnhammer, E. L. L. (2000) "The Pfam protein families database." *Nucleic Acids Research* **28**, 263-266.

Berman, H. M., Westbrook, J., Feng, Z., Gilliland, G., Bhat, T. N., Weissig, H., Shindyalov, I. N. and Bourne, P. E. (2000) "The Protein Data Bank." *Nucleic Acids Res.* **28**, 235-242.

Dayhoff, M. O. (1965-1978) *Atlas of Protein Sequence and Structure.* Volumes 1-5, Supplements 1-3. National Biomedical Research Foundation, Washington, DC.

Davidson, S. B., Overton, C. and Nuneman, P. (1995) "Challenges in integrating biological data sources." *Journal of Computational Biology* **2**, 557-572.

Eddy, S. R., Mitchison, G. and Durbin, R. (1995) "Maximum discrimination hidden Markov models of sequence consensus." *Journal of Computational Biology* **2**, 9-23.

Garavelli, J.S., Hou, Z., Pattabiraman, N. and Stephens, R. M. (2001) "The RESID database of protein structure modifications and the NRL-3D sequence-structure database." *Nucleic Acids Research* **29**, 199-201.

Hofmann, K., Bucher, P., Falquet, L. and Bairoch, A. (1999) "The PROSITE database, its status in 1999." *Nucleic Acids Research* **27**, 215-219.

Huang, H., Xiao, C. and Wu, C. H. (2000) "ProClass protein family database." *Nucleic Acids Research* **28**, 273-276.

Kanehisa, M. and Goto, S. (2000) "KEGG: Kyoto encyclopedia of genes and genomes." *Nucleic Acids Research* **28**, 27-30.

Karp, P. D. (1995) "A strategy for database interoperation." *Journal of Computational Biology* **2**, 573-586.

Laskowski, R. A. (2001) "PDBsum: Summaries and analyses of PDB structures." *Nucleic Acids Research* **29**, 221-222.

Lo Conte, L., Ailey, B., Hubbard, T. J. P., Brenner, S. E., Murzin, A. G. and Chothia, C. (2000) "SCOP: A structural classification of proteins database." *Nucleic Acids Research* **27**, 254-256.

Marcotte, E. M., Pellegrini, M., Thompson, M. J., Yeates, T. O. and Eisenberg, D. (1999) "A combined algorithm for genome-wide prediction of protein function." *Nature* **402**, 83-86.

Mewes, H. W., Frishman, D., Gruber, C., Geier, B., Haase, D., Kaps, A., Lemcke, K., Mannhaupt, G., Pfeiffer, F., Schuller, C., Stocker, S. and Weil, B. (2000) "MIPS: A database for genomes and protein sequences." *Nucleic Acids Research* **28**, 37-40.

McGarvey, P., Huang, H., Barker, W. C., Orcutt, B. C. and Wu, C. H. (2000) "The PIR Web site: New resource for bioinformatics." *Bioinformatics* **16**, 290-291.

McPherson, J. D., Marra, M., Hillier, L., Waterston, R. H., Chinwalla, A., *et al.* The International Human Genome Mapping Consortium (2001) "A physical map of the human genome." *Nature* **409**, 934-941.

Overbeek, R., Larsen, N., Pusch, G. D., D'Souza, M., Selkov, E. Jr., Kyrpides, N., Fonstein, M., Maltsev, N. and Selkov, E. (2000) "WIT: Integrated system for high-throughput genome sequence analysis and metabolic reconstruction." *Nucleic Acids Research* **28**, 123-125.

Pearl, F. M. G., Martin, N., Bray, J. E., Buchan, D. W., Harrison, A. P., Lee, D., Reeves, G. A., Shepherd, A. J., Sillitoe, I., Todd, A. E., Thornton, J. M. and Orengo, C. A. (2001) "A rapid classification protocol for the CATH domain database to support structural genomics." *Nucleic Acids Research* **29**, 223-227.

Pearson, W. R. and Lipman, D. J. (1988) "Improved tools for biological sequence comparision." *Proceedings of the National Academy of Sciences (USA)* **85**, 2444-2448.

Peterson, J. D., Umayam, L. A., Dickinson, T., Hickey, E. K. and White, O. (2001) "The comprehensive microbial resource." *Nucleic Acids Research* **29**, 123-125.

Pruitt, K.D. and Maglott, D. R. (2001) "RefSeq and LocusLink: NCBI gene-centered resources." *Nucleic Acids Research* **29**, 137-140.

Smith, T. F. and Waterman, M. S. (1981) "Comparison of bio-sequences." *Adv. Appl. Math.* **2**, 482-489.

Tatusov, R. L., Natale, D. A., Garkavtsev, I. V., Tatusova, T. A., Shankavaram, U. T., Rao, B. S., Kiryutin, B., Galperin, M. Y., Fedorova, N. D. and Koonin, E. V. (2001) "The COG database: New developments in phylogenetic classification of proteins from complete genomes." *Nucleic Acids Research* **29**, 22-28.

Thompson, J. D., Higgins, D. G. and Gibson, T. J. (1994) "CLUSTAL W: Improving the sensitivity of progressive multiple sequence alignment through sequence weighting, position-specific gap penalties and weight matrix choice." *Nucleic Acids Research* **22**, 4673-4680.

Venter, J. C., Adams, M. D., Myers, E. W., Li, P. W., Mural, R. J., *et al.*, The Celera Genomics Sequencing Team (2001) "The sequence of the human genome." *Science* **291**, 1304-1351.

Wheeler, D. L., Church, D. M., Lash, A. E., Leipe, D. D., Madden, T. L., Pontius, J. U., Schuler, G. D., Schriml, L. M., Tatusova, T. A., Wagner, L. and Rapp, B. A. (2001) "Database resources of the national center for biotechnology information." *Nucleic Acids Research* **29**, 11-16.

Wu, C. H., Huang, H., Arminski, L., Castro-Alvear, J., Chen, Y., Hu, Z., Ledley, R. S., Lewis, K. C., Mewes, H-W., Orcutt, B. C., Suzek, B. E., Tsugita, A., Vinayaka, C. R., Yeh, L-S., Zhang, J. and Barker, W. C. (2002) "The protein information resource: An integrated public resource of functional annotation of proteins." *Nucleic Acids Research* **30**, in press.

Wu, C. H., Huang, H. and McLarty, J. (1999) "Gene family identification network design for protein sequence analysis." *International Journal of Artificial Intelligence Tools* (Special Issue on Biocomputing) **8**, 419-432.

Wu, C. H., Xiao, C., Hou, Z., Huang, H. and Barker, W. C. (2001) "iProClass: An integrated, comprehensive, and annotated protein classification database." *Nucleic Acids Research* **29**, 52-54.

Wu, C. H., Zhao, S. and Chen, H. L. (1996) "A protein class database organized with ProSite protein groups and PIR superfamilies." *Journal of Computational Biology* **3**, 547-562.

Xenarios, I., Fernandez, E., Salwinski, L., Duan, X. J., Thompson, M. J., Marcotte, E. M. and Eisenberg, D. (2001) "DIP: The database of interacting proteins: 2001 update." *Nucleic Acids Research* **29**, 239-241.

Yoshida, M., Fukuda, K. and Takagi, T. (2000) "PNAD-CSS: A workbench for constructing a protein name abbreviation dictionary." *Bioinformatics* **16**, 169-175.

Author's Address

Cathy H. Wu, National Biomedical Research Foundation, Georgetown University Medical Center, 3900 Reservoir Road, NW, Washington, DC 20007-2195, USA. Email: wuc@georgetown.edu.

Chapter 6

High-Grade Ore for Data Mining in 3D Structures

Jane S. Richardson and David C. Richardson

6.1 Introduction

The 3-dimensional structures of proteins, nucleic acids, and complexes are becoming an increasingly important part of bioinformatics, with the advent of structural genomics, protein-protein interaction analysis, and large-scale functional genomics. Although not approaching the extent of sequence data, the 3D database is huge, complex, and growing rapidly. The quality of 3D data is very good but varies increasingly widely as more protein crystal structures are being done at atomic resolution, while ever-larger molecular machinery can be successfully tackled at meaningful but quite low resolution. NMR (nuclear magnetic resonance) methodology for determining 3D structures is currently changing even more rapidly than crystallography. As in any data analysis effort, an understanding and assessment of possible errors is a crucial aspect of doing structural bioinformatics.

After a brief review of the traditional criteria for assessing structural quality, this chapter will concentrate on a new validation method that is especially powerful yet quite accessible to non-experts. This method, called all-atom contact analysis, is also applicable to evaluating the ligand-protein, protein-protein, or protein-nucleic acid interactions seen in complexes.

The series of illustrations in Figure 1 a-f spans for one particular example the transition from 1D sequence information to 3D structure, in increasingly detailed representations, and then to all-atom contact information. The sequence in Figure 1a flags the 14 residues that are conserved across the ribonuclease-A-like superfamily from a clustal W alignment (http://pir.georgetown.edu). When that sequence is visualized on the 3D structure, it becomes apparent that the alignment missed 2 conserved residues (Phe8 and His12) because of an unrecognized gap in one subfamily of the proteins. It is also clear that the 16 conserved residues divide into two groups: one group of 5 residues exposed at the active site cleft (starred in Figure 1b) and another group of 11 structurally important residues in the two hydrophobic cores, including three of the four disulfide bridges. The overall

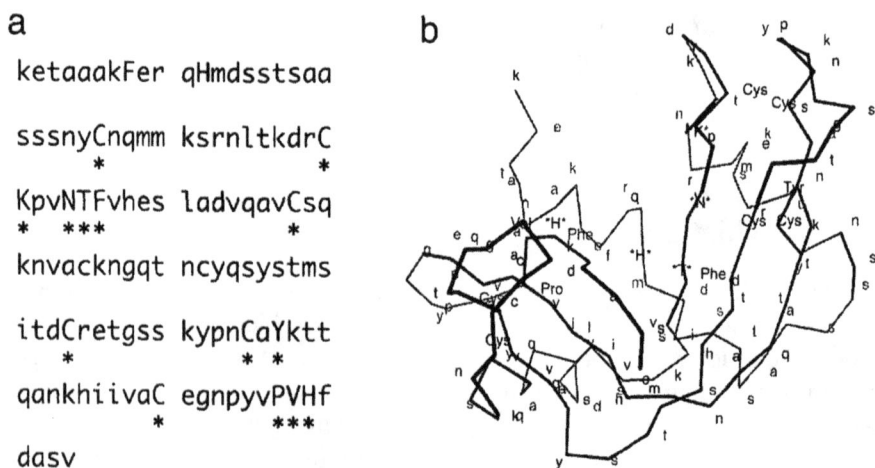

Figure 1. The progression of molecular information from sequence to 3D fold to structural details, illustrated using the structure of bovine ribonuclease A at 1.26Å resolution (PDB file 7RSA [Wlodawer *et al.*, 1988]). a) Sequence; the 14 residues shown as conserved across the ribonuclease-A-like superfamily by Clustal-W alignment are starred. b) Cα backbone, with sequence arranged in 3D on the molecule; the 5 conserved residues in the active site cleft (H,K,N,T,H) are flanked by stars, and the 11 structurally-conserved residues of the two domain cores are labelled in three-letter code.

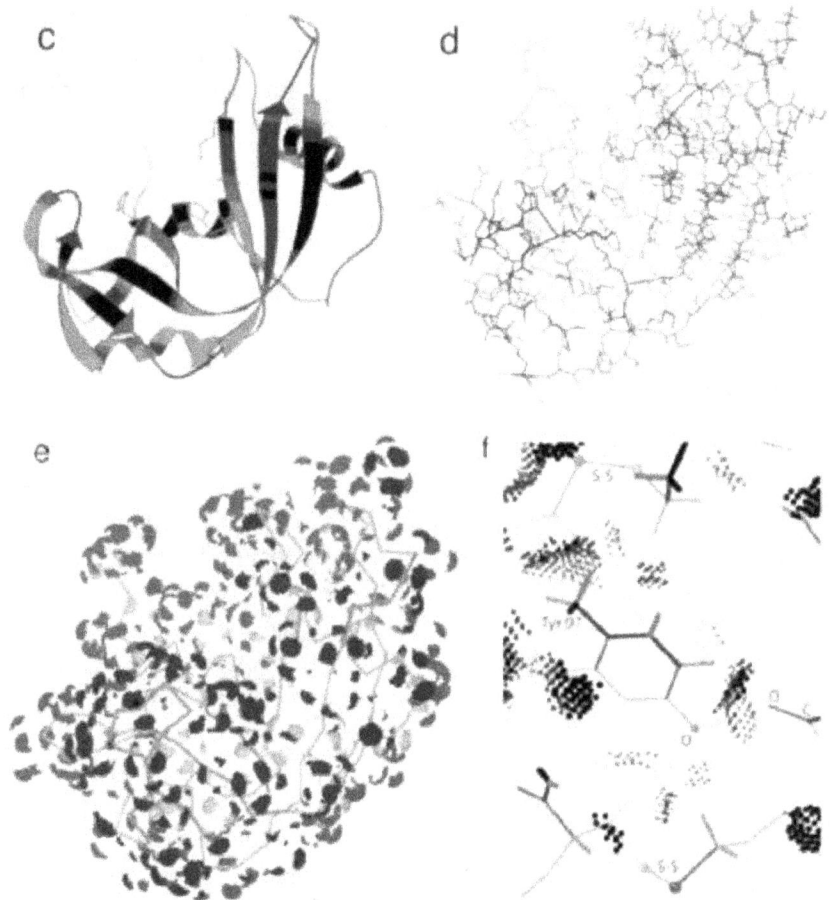

Figure 1 cont'd. c) Backbone ribbon schematic, with active site sidechains and secondary structure (α-helices are spiral ribbons and β-strands are arrows). d) Stick figure, with all non-H atoms in the left half and all atoms including hydrogens on the right; the active site is starred. e) Dots outlining the solvent-accessible parts of the van der Waals surface, for just the H atoms, which make up 53% of the outer surface. f) All-atom contact dots around the structurally-conserved Tyr97 sidechain, showing the OH H-bond to backbone and the ring van der Waals contacts with the two disulfides above and below it.

fold, with its evolutionary implications, is best recognized in backbone or ribbon representations, as in Figure 1c, although the interactions determining that fold involve all of the atoms, including the often-ignored hydrogens (Figure 1d). The protein surface, over half of which is formed by H atoms (Figure 1e), provides the biologically significant interactions with substrates, inhibitors, and other macromolecules. All-atom contact analysis provides a way of both visualizing and quantifying geometrical goodness-of-fit, for study of internal packing or of interactions between molecules; in Figure 1f, the contact dots show why Tyr97 is especially suited to stabilizing the core of the righthand lobe of ribonuclease.

6.2 Traditional Quality Measures

Measures of structure quality can apply either overall or locally, and some measures do both. For NMR structures, the most important overall criterion is the number of measured restraints (such as NOE restraints on atom-atom distances, or J-coupling restraints on torsion angles); very well-determined structures might have 20, or even 40 restraints per residue. Another criterion is the RMS deviation among the multiple models, or the closeness of overlap in a figure showing the entire ensemble. Divergence of the models is especially important for identifying regions that either are less well determined or are actually mobile in the molecule.

For crystal structures, the most important overall criterion of accuracy is the resolution: at 3Å resolution the fold and secondary structure can be determined; at 2Å resolution the detailed backbone and sidechain conformations are generally reliable; at 1Å resolution atoms can be seen as individual balls of density centered perhaps within 0.1Å of true position, and multiple conformations and water structure can be distinguished. The residual, or R-factor, measures agreement between observed and model-calculated diffraction data, while the "free R" measures agreement with a subset of the observed data deliberately left out of the refinement process for an unbiased evaluation [Brunger, 1992]. At 2Å resolution a good structure should have an R of about 20% or less and a free R perhaps 4-5% higher. Resolution, R, and free R are reported in parseable form in the header of a PDB (Protein Data Bank; [Berman *et al.*, 2000]) file, so they are readily available whenever a 3D coordinate file is consulted.

Another very useful criterion of overall quality is the Ramachandran plot (a 2D plot of the ϕ,ψ dihedral angles that define backbone conformation),

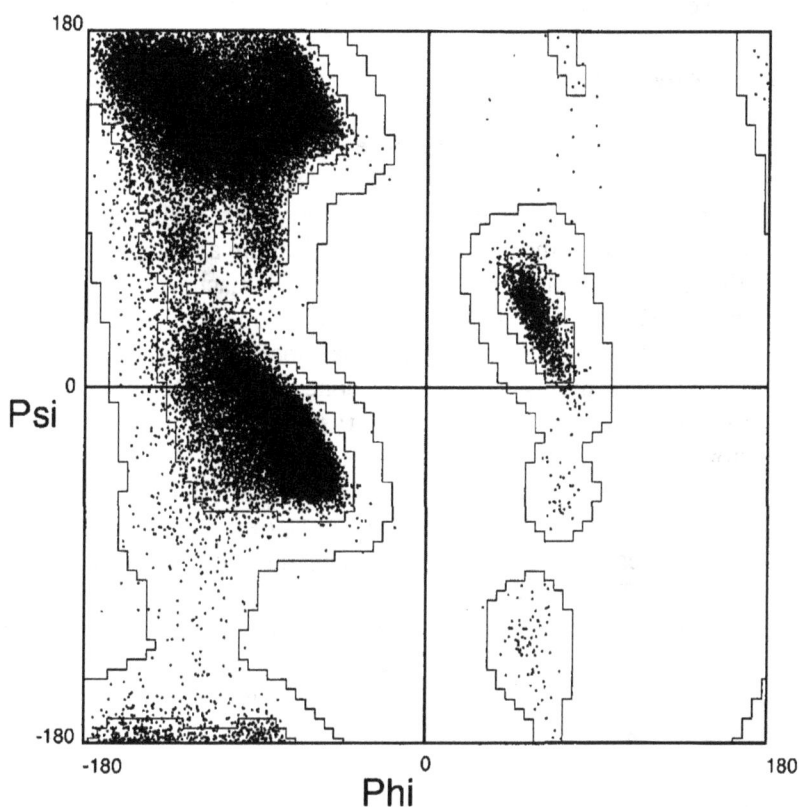

Figure 2. The Ramachandran plot of φ versus ψ backbone conformational angles, used as a structure validation tool. The points plot the individual φ,ψ values for each well-ordered residue with a crystallographic *B*-factor <30 in 500 protein structures at 1.8Å resolution or better [RichardsonLabWebSite, 2001]. The inner outline encloses the preferred "core" region containing 98% of this high-quality data, and the outer line encloses the region of allowed but somewhat strained conformations.

produced by "validation" software (available on the web) such as ProCheck [Laskowski *et al.*, 1993] and WhatIf [Vriend, 1990]. Preferred, core regions of the plot have been defined empirically, and almost all φ,ψ values in a structure should fall within those regions, as shown in Figure 2. The validation programs will produce a Ramachandran plot for a submitted structure, give the percentage within core regions, and compare that number to the scores expected at different resolutions. In the absolute sense, about 98% of non-Gly protein residues actually have core ϕ,ψ values; 2% of residues genuinely occupy slightly strained conformations, while any excess of non-Gly outliers above 2% represent errors. Like the free R value, the Ramachandran criterion has the advantage that it evaluates a quantity not directly optimized by structure refinement; this gives it more sensitivity than the ordinary R-factor or the ideality of bond lengths and angles, which are part of the target function for refinement. Ramachandran criteria are applicable to NMR as well as x-ray protein structures, but the equivalent criteria are not yet available for nucleic acids, since the database is still much too small to deal with the six variable backbone angles in RNA or DNA.

The most important local criterion of crystallographic structure quality is the B-factor, or temperature factor, which is a measure of the sharpness vs. broadness of the electron density seen at each atom. High B-factors can be caused by motion or static disorder of the atoms, or can result from errors in the data, the phases, or the fitting and refinement; regardless of cause, the position of a high-B atom is known with less precision. Average B-factors vary somewhat with refinement strategy and especially with resolution (the electron density is necessarily broader at low resolution). The most important differences in B, however, are within a given structure: most parts will be clear and well-ordered with low B's, while some parts (usually some of the loops, ends, or sidechains on the outside) may be quite disordered and have very high B's. Sometimes no density at all is visible for such atoms, in which case either their coordinates may be left out altogether or else their B-factors will refine to the maximum allowed in order to smear them into appropriate invisibility. Therefore, even in a high-resolution structure it is important to look at the B-factors for any region of particular interest. This is easy to do, since the B is given as the last numerical field on every atom record in the PDB coordinate file. In general, the highest-B regions of a given structure are probably unreliable, and as a rule of thumb at 2Å resolution, atoms with B>40 are suspect. Most high-B sidechains are in fact correctly placed, but they are more of a gamble: one with a B of 50 is ten times more likely to have the wrong conformation than one with a B of 10-20 [Word *et al.*, 1999a]. For NMR structures, the nearest analog of the B-factor is the local amount of deviation among models, easily judged by viewing the ensemble of models superimposed.

In a high-resolution crystal structure, many of the regions with partial disorder can be seen and fitted as alternate conformations (usually just two,

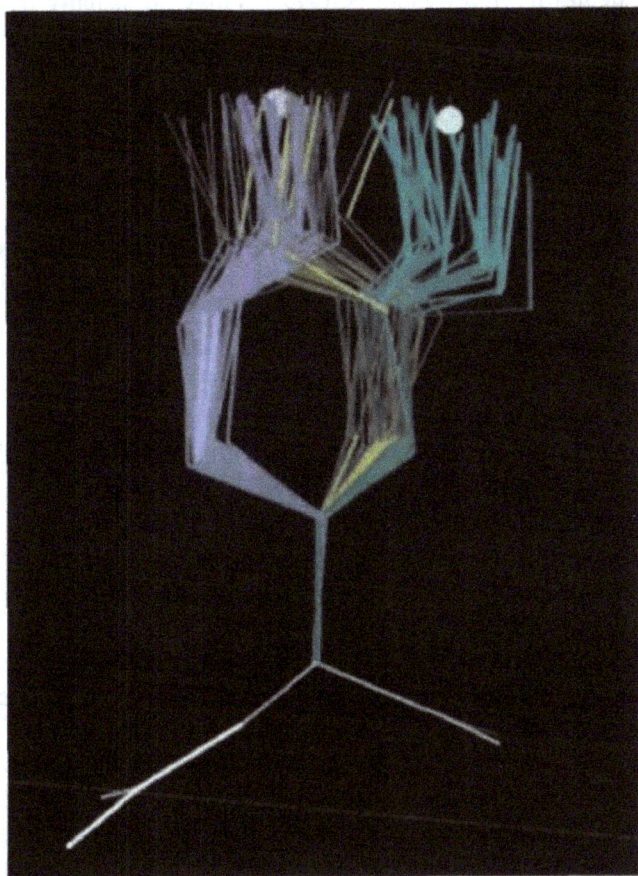

Figure 3. Three neighboring sidechain rotamers for lysine. All examples with sidechain χ angles within ±60° of these rotamers were superimposed, for Lys sidechains with *B*-factor <40 in a dataset of 240 protein structures at 1.7Å resolution or better [Lovell *et al.*, 2000]. Each rotamer is a different color, and the balls mark the average Nζ position for each rotamer. Even for the long, flexible lysine sidechain, conformations are clearly distinct and quite tightly clustered.

marked a and b). Such alternate conformations can represent the local structure better than a single conformation with a high B-factor; however, they still are less accurate than well-ordered single conformations, because such overlapping density is difficult to fit correctly. On average, the "b" conformations are more error-prone, because they are usually the ones with lower occupancy and also because validation programs usually do not check the "b" geometry.

The second traditional criterion of local accuracy is the deviation of bond angles from ideal values. (Bond length deviations are not very useful for assessment purposes, because their values are very tightly restrained.) Those angle deviations are tabulated and summarized by the standard validation programs. Genuine angle variations of 1 or 2 standard deviations, or about 2-4°, are relatively common, and a bit more than that is occasionally orchestrated by the protein where needed at an active site. Bond angles are often quite deviant at the junction between single and alternate conformations, but those are basically technical glitches and do not meaningfully signal whether or not there are underlying problems with the model. However, in our experience a deviation of >5σ from ideal in any angle within a single-conformation region almost always indicates an incorrect local conformation. Bond angles at the Cα are especially diagnostic, since that is where sidechain and backbone meet. Some such cases will be discussed below, because they are also often signaled by all-atom clashes.

The third local criterion is the degree to which sidechain torsion angles match one of the preferred sets of values called a sidechain rotamer. In spite of both positive and negative interactions in the folded protein, sidechains adopt a discrete rather than a continuous distribution of conformations, as illustrated for a sub-population of lysines in Figure 3. An up-to-date library of the possible rotamers and an extensive discussion of the issues involved can be found in [Lovell *et al.*, 2000]. For the purposes of spotting potential problem areas in a structure, a very simplified but useful starting-point is to watch out for eclipsed torsion angles around bonds with tetrahedral geometry at both ends. In particular, sidechains branched at Cβ (Thr, Val, or Ile) are almost certainly incorrect if they have an eclipsed χ_1 angle.

6.3 All-Atom Contacts for Assessing Structural Accuracy

Recently we discovered a powerful new source of independent information for structure validation in an unexpected place: the hydrogen atoms. Although crystal structures can be solved successfully and accurately without the H atoms and NMR structures can be solved well without treating hydrogens at 100% radius, their inclusion can help enormously. Nearly half the atoms in biological macromolecules are hydrogens. Their positions are mostly constrained by the geometry of the other atoms, and their packing interactions are extremely demanding. Any significant error on the inside of a structure shows up very clearly in the form of physically impossible overlaps of H atoms with each other or with the other atoms. An especially important aspect of all-atom contact analysis is that in addition to locating problems it can very often suggest how to fix them.

The first essential step in all-atom contact analysis is obviously to add and optimize the H atoms, which is done by a program called Reduce [Word *et al.*, 1999b]. Most nonpolar and some polar H positions are determined by the heavier atoms and can simply be added in standard geometry (e.g., methylenes, aromatic H, peptide NH, etc.). At the other extreme, OH positions can rotate quite freely and must be optimized relative to their surroundings. We have determined from very high-resolution structures and neutron diffraction data that NH_3 groups and the terminal methyls of Met sidechains can adopt equilibrium orientations significantly away from staggered and must therefore be optimized, but the equilibrium orientations of other methyls are remarkably well relaxed and can be satisfactorily treated as staggered. The sidechain amides of Asn or Gln and the ring orientation of His are fairly often misassigned by 180°, since it is difficult to distinguish the N vs O or N vs C atoms in electron density maps; therefore Reduce considers possible Asn/Gln/His flips when placing hydrogens. Finally, these optimizations must be done jointly in the context of entire local H-bond networks. Reduce handles proteins, nucleic acids, and small-molecule ligands and produces a commented output file in PDB format. Like all our software, it is freely available on our web site [RichardsonLabWebSite, 2001].

The second step is using the all-atom coordinate file to calculate the favorable and unfavorable contacts between the atoms. This is done by the program Probe [Word *et al.*, 1999a], which calculates all-atom contacts for display or quantification, using an algorithm illustrated schematically in Figure 4. A small spherical probe 0.25Å in radius is rolled over the surface of each atom, and a dot is generated only if the probe sphere touches another not-covalently-bonded atom. Favorable van der Waals contacts, with a minimum of 0Å and a maximum of 0.5Å gap between the atom surfaces, are shown as surface patches of dots color-coded by gap size. Favorable overlaps between H-bond donor and acceptor atoms are shown as pale green

Figure 4. Calculation and display of all-atom contacts illustrated on a thin slice through a small piece of protein structure. Fine gray dots (here for didactic purposes but normally not shown) show the van der Waals surfaces of all the atoms including hydrogens. A small spherical probe (gray ball) just 0.25Å in diameter is rolled over the surface of each atom, leaving a color-coded contact dot wherever the probe also touches or intersects another atom not within three covalent bonds of the first. As labelled in the figure, the resulting pairs of contact patches come in three types. Favorable van der Waals contacts are shown by dots that are blue when the local gap between atomic surfaces is near the 0.5Å maximum, shading to green as the gap aproaches zero. The favorable overlaps of suitable donor and acceptor atoms that constitute H-bonds are shown by pale green dots, forming lens or pillow shapes. Unfavorable overlaps of all other, non-compatible, atom pairs are emphasized with "spikes" rather than dots, color-coded from yellow for the slight overlaps that still represent good contacts through to red and hotpink for the physically-impossible atomic overlaps that cannot occur in the real molecule and must represent model errors.

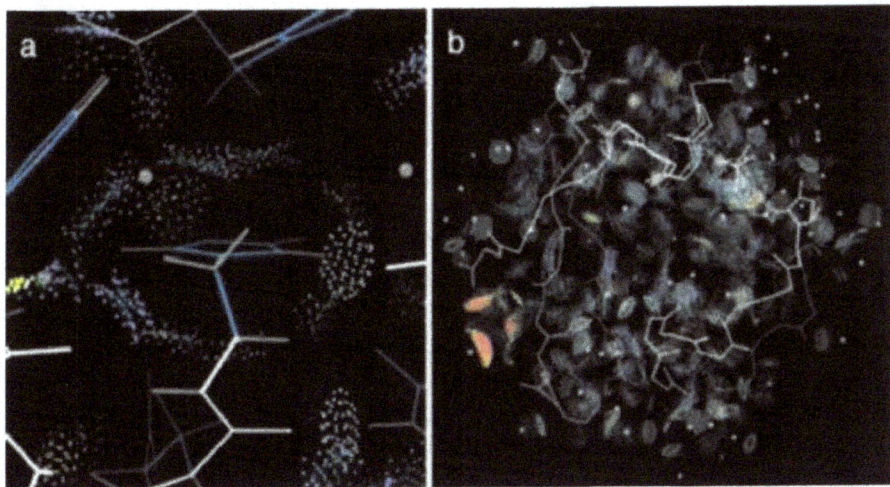

Figure 5. Examples of all-atom contacts for well-packed, accurate structures. a) Contacts around active-site His12 in the 7RSA ribonuclease, with H-bonds (pale green dots) for each ring N and dense, well-fitted van der Waals contacts to surrounding backbone and sidechains. The pale orange balls are waters. b) All-atom contacts for the entire molecule of the 1RB9 rubredoxin structure at 0.92Å resolution [Dauter *et al.*, 1999], showing the blues and greens of dense, well-fitted packing throughout except for one apparently misfit Lys sidechain at the left, which has three serious clashes, poor χ angles, and high *B*-factors.

dots, forming lens or pillow shapes. Unfavorable atomic overlaps, or "clashes", are emphasized by spikes rather than dots: yellow for slight overlaps, shading to bright red and hot pink for the worst cases. Interpreting these displays is thus very easy: red spikes are bad, while lots of cool green is good; for example, Figure 5a shows the excellent contacts around His12 in the 7RSA ribonuclease.

For numerically analyzing goodness-of-fit inside or between the molecules in a model, the following 3-term Probe score combines the contact, H-bond, and clash components:

$$contactscore = \sum_{dots} e^{-(gap/err)^2} + 4 \times Vol(Hbonds) - 10 \times Vol(overlaps)$$

This "contact score" is formulated geometrically rather than energetically, because a bad overlap means a mistake in the model, not a high energy. A good contact score is positive rather than negative, to emphasize that distinction. For assessing crystallographic model quality, the simpler "clashscore" is just the number of clash overlaps >0.4Å, normalized per 1000 atoms (a good clashscore is low). As expected for a measure of structure quality, the overall clashscore is highly correlated with resolution, and the local clashscore is even more strongly correlated with B-factor [Word *et al.*, 1999a]. Contact score can be directly calculated by Probe, and a Unix script called Clashlistcluster is available that gives both clashscore and a spatially-clustered list of the serious clashes in a structure.

The clashscore is not as easily applied to validate NMR structures, because H-H distances are directly used in NMR refinement and with some methodologies the clashes can be avoided simply by expanding the entire model somewhat (indicated by a sparseness of contacts and H-bonds). However, most NMR structures show clashes in their problem areas, and a clash list or display is very useful for finding and fixing those problems.

The most powerful form of all-atom contact analysis is actually the visual display, because the local pattern of clashes and contacts often suggests the origin and solution of the problem. Such display is most easily done in our Mage display program [Richardson and Richardson, 1992; Richardson and Richardson, 2001], which allows turning on or off the different contact types and choosing to color the dots by gap size or by atom type. In Probe's default mode all internal sidechain-sidechain and sidechain-backbone contacts, plus contacts with ligands and waters, are calculated for the entire molecule and can then be displayed and explored. (Note, however, that for nucleic acids it is vital to calculate backbone-backbone contacts as well.) Such a contact display is shown in Figure 5b for the 1RB9 rubredoxin [Dauter *et al.*, 1999] at 0.92Å resolution. As typical of most structures at atomic resolution, almost all of the model fits the all-atom criteria beautifully, with no bad clashes, no flips needed, and dense, well-fitted packing shown by the green dot patches. Also quite typically, there is one isolated region with a set of serious clashes; in this case the problem was apparently caused by fitting a Lys sidechain into what should have been water density and vice-versa, losing potential H-bonds and necessitating a very poor Lys rotamer.

Even in very excellent structures, therefore, it is worth watching out for the few isolated problems, while at more modest resolutions there are almost always clashes indicating errors that could be fixed by the depositors or that can flag locally unreliable regions for the bioinformatician. Equally importantly, a well-packed region with extensive green contacts and no red spikes can validate the accuracy of that region even at lower resolutions.

6.4 Patterns of Common Misfittings

As mentioned above, the 180° flip state of Asn, Gln, and His sidechains are easy to misassign in crystallographic structures due to the ambiguity of atom type in electron density maps. Analysis of H-bonding can determine a majority of these cases, and if potential clashes of the larger NH_2 group are also considered, then flip assignments become blatant rather than subtle, except for about 15% of these residues on the protein surface that almost certainly adopt both conformations [Word *et al.*, 1999b]. Figure 6a and 6b show an Asn-Gln-Asn H-bond network in 7RSA [Wlodawer *et al.*, 1988], in its best arrangement (as deposited; part a) and again with each amide flipped and all donors and acceptors interchanged (part b). As found for all cases where both H-bond and clash aspects are unambiguous, the two criteria agree on the answer: here 8 rather than 4 H-bonds are clearly preferable, and zero rather than 3 serious clashes are clearly preferable. If considered in isolation, the central Gln would only be fairly weakly determined by H-bond geometry to the water, but in combination with the two flanking Asn the entire network is determined overwhelmingly. Occasionally, even buried Asn/Gln make no H-bonds, but the orientation can then be determined by clashes of the NH_2 group in the incorrect flip state.

Amide flip problems should not occur in NMR structures if the NH_2 protons were assigned and they had NOEs. If not, then the orientation may well be wrong, but not necessarily by 180°. For crystal structures, the Asn/Gln/His flip may be automatically corrected by Reduce in its "-build" mode, producing a modified and commented PDB file in standard format with H atoms added and flips corrected. Since we do not believe that even the best automatic algorithms should be accepted without scrutiny, a script called Flipkin is available for analyzing Reduce's decisions: it produces a kinemage file for display in Mage, with a view for each Asn/Gln/His sidechain that animates between contacts for the two best flip states. The comparison pairs in Figure 6 were taken from the flip kinemage for 7RSA.

Sidechains with tetrahedral geometry, somewhat surprisingly, are also fairly often fit backwards. In NMR, this happens because the stereospecific assignment of resonances to the methyls of the sidechain branch of a Val or Leu either could not be done or were erroneous. In crystallography, it

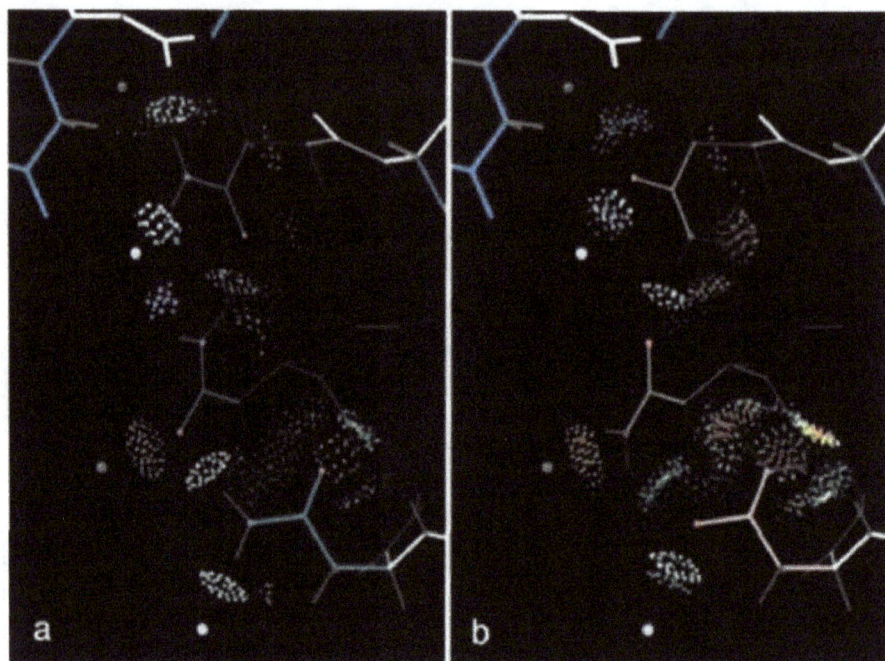

Figure 6. Assigning amide flip orientations within an H-bond network around Gln69 in the 7RSA ribonuclease structure. a) The best combination of orientations (in green, as found in 7RSA), with 8 H-bonds and no clashes. b) The combination with all amides flipped (in pink), which keeps 4 H-bonds but is ruled out by bad clashes of all three NH₂ groups.

happens because less-than-optimal electron density for a tetrahedral branch can often appear straight across rather than showing the tetrahedral angle, so that either fitting looks equally plausible as in the lower half of Figure 7a.

Although 7RSA has no problems of this sort, it corrects two Thr that were fit backwards in the earlier 5RSA [Wlodawer *et al.*, 1986] structure at 2Å resolution.   In Figure 7b the two conformations for Thr 87 are superimposed, showing that angles around the Cα and Cβ had to be strongly distorted in order for refinement to get the bulk of the sidechain approximately into density. The misfit Thr also has an eclipsed $\chi 1$ angle. Figure 8a shows the diagnostic clashes produced by the misfit conformation in the deposited structure, while Figure 8b shows that the problem could have

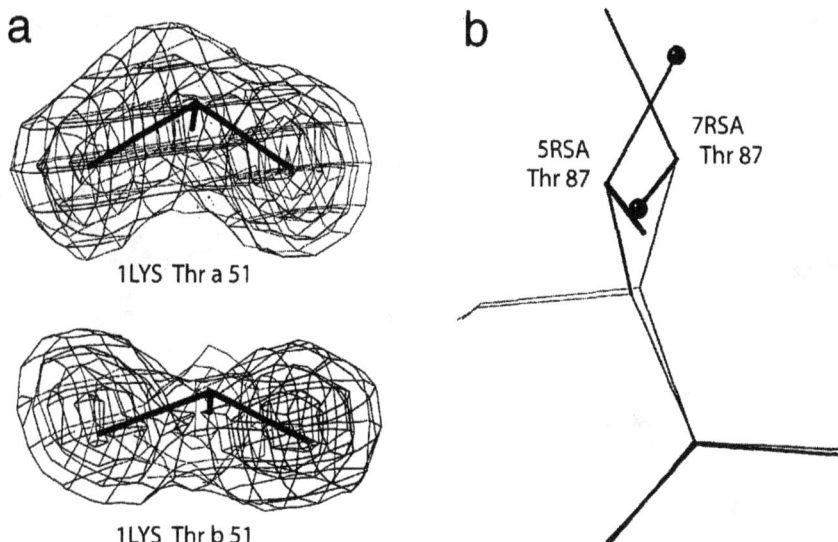

Figure 7. a) Unambiguous electron density for a Thr sidechain that clearly shows the Cβ position (above) contrasted with ambiguous electron density (below) that makes the eclipsed conformation (as shown) appear to be just as good a match as the correct one (rotated 180°); from PDB file 1LYS [Harata, 1994]. b) Such a misfit Thr sidechain (from 5RSA at 2.0Å resolution [Wlodawer *et al.*, 1986]), superimposed on the corrected version (from 7RSA at 1.26Å resolution); the misfit Cβ position was shifted by 0.7Å in order to force the Cγ and Oγ into density.

been successfully corrected by using our tools on the 5RSA structure. On Unix or Linux, Mage has the capability of idealizing sidechain geometry and interactively updating the all-atom contact display as rotamers are tried or individual torsion angles are adjusted [Word *et al.*, 2000]. Both Cβ-branched sidechains (Thr, Val, Ile) and longer sidechains such as Leu or Met can have problems of this sort; such cases can often be located by their deviant bond angles or torsion angles, but they can nearly always be found by their severe all-atom clashes.

A third common type of misfitting involves cases, usually in backbone, where there are too many variable conformational angles per observable. Glycine residues in proteins are more error-prone than other amino acids

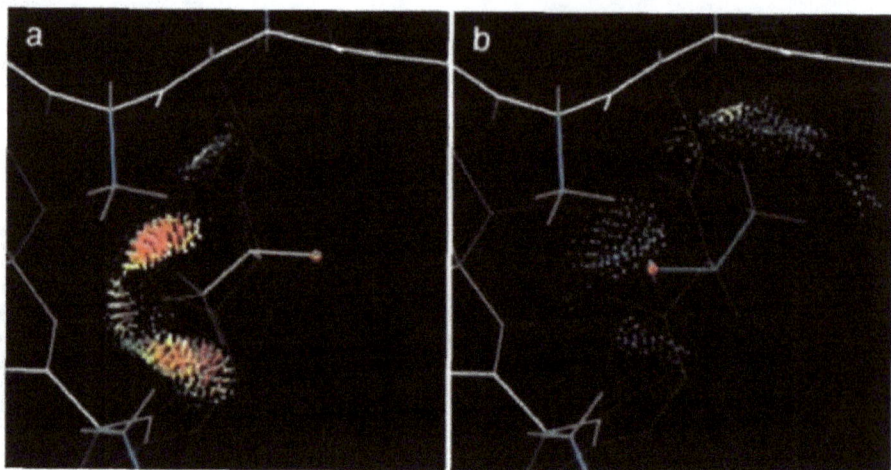

Figure 8. How the interactive all-atom contact tools could have corrected the misfit Thr in Figure 7b, using only the original structure. a) All-atom contacts for Thr 87 in 5RSA, with eclipsed χ_1, bad bond angles, and serious clashes around the Cγ methyl. b) Ideal-geometry Thr rotamers were tried and adjusted in the interactive MAGE/PROBE system [Word *et al.*, 2000], producing the excellent fit shown, with a slightly long H-bond. Further refinement would probably move the backbone slightly, but one could be sure this is close to the right answer, confirmed by the 7RSA structure.

because there is no observable Cβ to help show the orientation; an example is discussed in [Word *et al.*, 1999a]. Nucleic acid backbone has an even more serious form of this same problem, for both x-ray and NMR structures, since there are six variable angles per residue along the sugar-phosphate backbone. The bases have large flat ring structures that are easy to locate accurately, and their all-atom contacts are almost always excellent in crystal structures, as for the tRNA example in Figure 9a. B-form DNA, and to a somewhat lesser extent A-form RNA, has been very well characterized structurally, so that the regular double-helical backbones generally show very good all-atom contacts. However, the less regular conformations, especially common and functionally important in RNA, show a high rate of physically impossible clashes such as the one in Figure 9b. Phosphates are readily positioned by their high, distinctive electron density; the approximate position of the backbone would seldom be wrong, but the detailed angles and

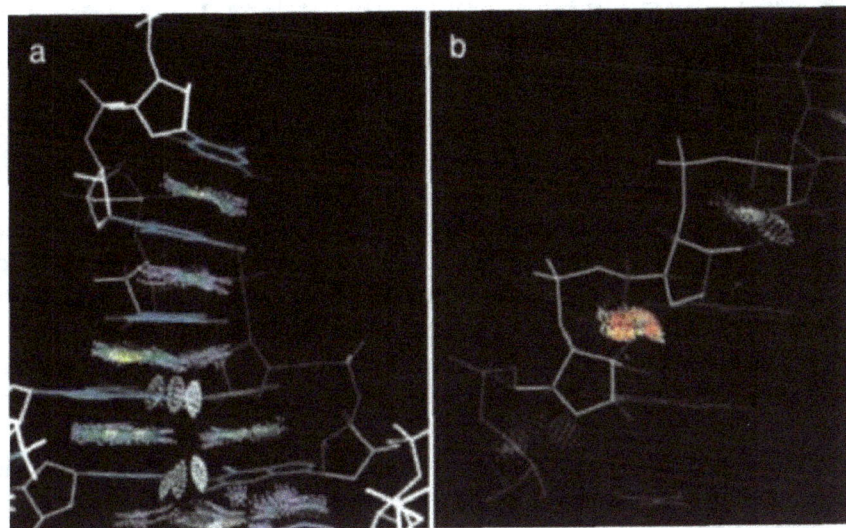

Figure 9. All-atom contacts in nucleic acid structures, illustrated from the 1EHZ tRNA at 1.93Å resolution [Shi and Moore, 2000]. a) Base-base contacts near the CCA end, showing base-pair H-bonds and very well-fitted base stacking interactions, both for successive base pairs and for single-base stacking. b) A local region with backbone-backbone all-atom contacts: one nucleotide residue shows a bad clash while the next residue has a closely similar conformation but favorable interactions. A backbone rotamer library might help improve RNA structures by replacing the first sort of conformation with the second sort.

orientations can often be incorrect. We hope in the future to provide tools to help increase the accuracy of RNA structures, such as a comprehensive backbone rotamer library containing only conformations free of serious clashes. In the meantime, however, the display or listing of all-atom clashes can be used to assess relative levels of local accuracy, especially when making functional comparisons among RNA molecules.

6.5 All-Atom Contacts for Characterizing Molecular Complexes

The calculation and display of all-atom contacts between two molecules can, of course, be effective in finding any problems with the 3D model at the interface. However, it also provides an intuitive but very detailed analysis of the specific atomic contacts between the molecules. As an example, Figure 10 shows the binding to ribonuclease A of a uridine vanadate inhibitor which mimics the 2',3' cyclic phosphate intermediate in catalysis. The contacts are, indeed, very tight and specific, with 7 strong H-bonds and much good van der Waals contact. Two stretches of mainchain and 7 sidechains, including the active site His12 and His119, interact with two-thirds of the inhibitor atoms.

Larger protein-protein complexes can also be studied effectively with these tools in the interactive mode, but the contacts are hard to show in a small static image. Figure 11 shows the complex of ribonuclease A (ribbon) with the large horseshoe-shaped α/β molecule of ribonuclease inhibitor (mainchain), with contact dots calculated between them. Ribonuclease is, indeed, held between the two ends of the horseshoe, but one loop also makes substantial contact with a Trp cluster on the inner face of the inhibitor. A sulfate is bound at the active site, between the enzyme and the inhibitor. Although the interface is extensive, its contacts are actually rather sparse as compared, for instance, with the very tight packing seen in Figure 10.

6.6 Discussion

When "validating" structures, it is important to keep in mind that not all problems can be resolved and that some errors are very significant while others really do not matter. A partially disordered surface loop genuinely does not have a single equilibrium structure but might in actuality be a mixture of three mostly-overlapping possible conformations each in equally favorable local energy minima. It may never be possible to disentangle them correctly in the 3D model, and there are very few purposes for which that would matter. Similarly, we are not concerned here with inaccuracies where, say, one or two torsion angles might be off by 20-30° but the conformation is in the right local minimum. Such inaccuracies typically change atom positions less than an Å, and usually they only produce clashes smaller than

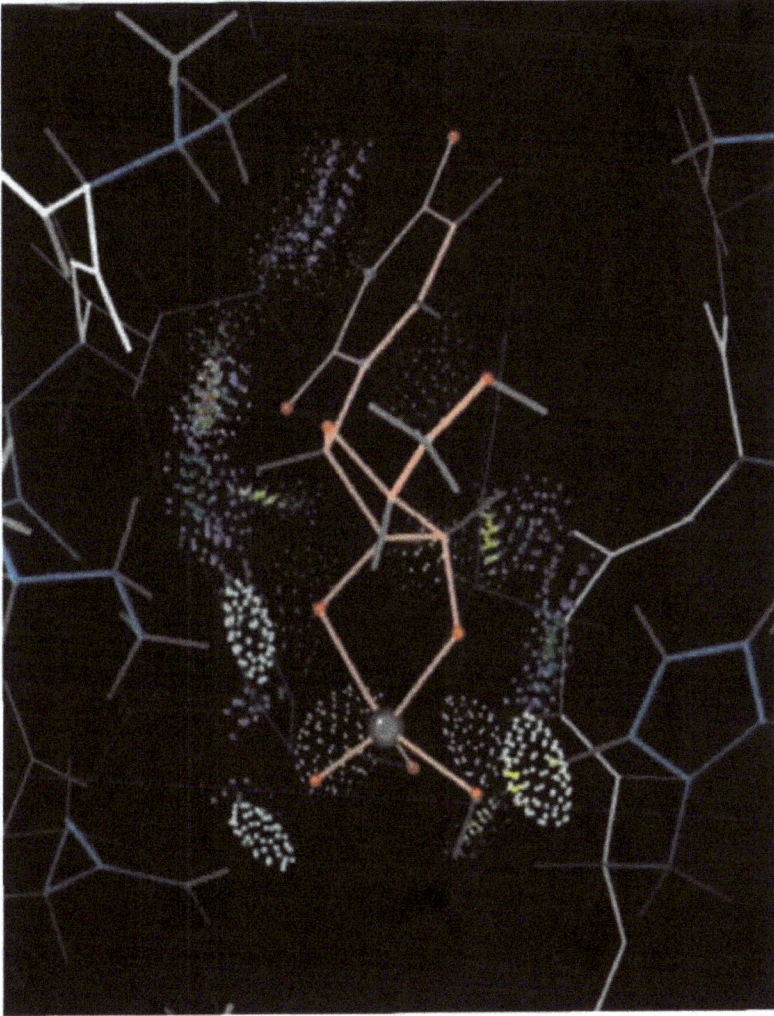

Figure 10. All-atom contacts between ribonucease A and the cyclic uridine vanadate intermediate-analog, in the 1RUV complex at 1.3Å resolution [Ladner *et al.*, 1997]. The ligand is nearly surrounded by interactions to both backbone and sidechains of the protein, with 7 H-bonds including those to the active-site His 12 and His 119 and extensive van der Waals contact.

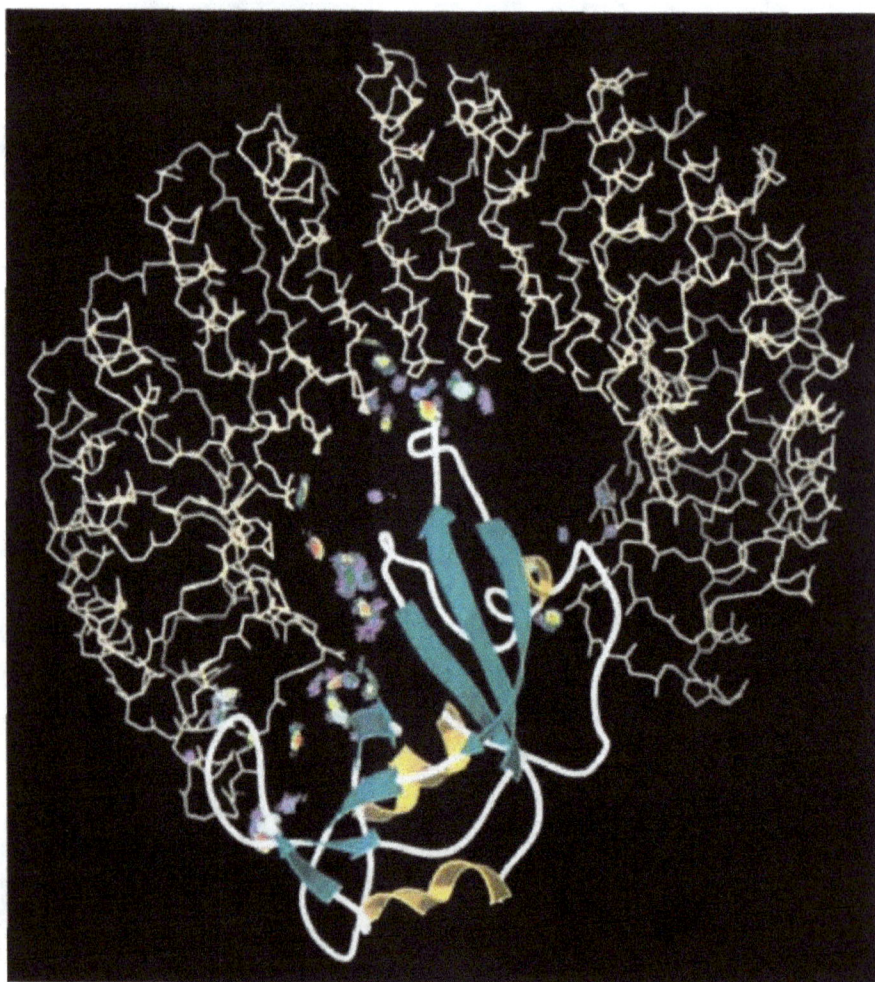

Figure 11. All-atom contacts between ribonuclease (ribbons) and the large α/β horseshoe of ribonuclease inhibitor (mainchain in peach) from the 1DFJ complex at 2.5Å resolution [Kobe and Deisenhofer, 1995]. The enzyme is bound between the two ends of the horseshoe, with the active site covered by the larger contact area at left, but there is also a significant contact of one loop with the inner face of the horseshoe. Although the contact surface at left is large, it is actually rather sparse, with about the same area of all-atom contact as the smaller but denser interaction of the uridine vanadate ligand in Figure 10.

our cutoff of 0.4Å. They would be crucial to a detailed chemical analysis of an enzymatic mechanism, but they would not change a bioinformatic analysis of fold, homology, function, or molecular interactions. When a local conformation is in the wrong energy well, however, such as a sidechain fitted backwards, that almost always does matter since it typically changes atom positions by 4 or 5Å. A flipped amide or His at an active site could make a functionally important H-bond network look absent; a backwards Leu in the core could seriously compromise homology modeling; and, any backwards sidechain at a binding site is likely to prevent successful ligand docking or drug design.

The application of all-atom contact analysis to the database of 3D biological structures has two quite distinct purposes. One is to provide simple, user-friendly tools that help anyone using the database to assess and thus take into account the relative accuracy of the structures, especially the local accuracy of any features of special interest. Our website provides a service called MolProbity that runs these tools on any user-designated file. The second purpose is to enable and encourage the use of these methods by structural biologists (see for instance [Richardson and Richardson, 2001] for a description of crystallographic tools), in order to correct as many problems as possible before structure deposition and thus to improve the 3D data directly. The latter aim would improve the grade of ore available for 3D data mining, while the former would improve the extraction process.

Acknowledgments

Thanks to Lizbeth Videau for doing web searches. Research on the all-atom contact method is supported by NIH grant GM-15000, and development of its use for structure validation and improvement is supported by NIH grant GM-61302.

References

Berman, H.M., Westbrook, J., Feng, Z., Gilliland, G., Bhat, T.N., Weissig, H., Shindyalov, I.N. and Bourne, P.E. (2000) "The Protein Data Bank." *Nucleic Acids Research* **28**, 235-242.

Brunger, A.T. (1992) "Free R-value — A novel statistical quantity for assessing the accuracy of crystal structures." *Nature* **355**, 472-475.

Dauter, Z., Butterworth, S., Sieker, L.C., Sheldrick, G. and Wilson, K.S. (1999) "Anisotropic refinement of rubredoxin from desulfovibrio vulgaris." To be Published.

Harata, K. (1994) "X-ray structure of a monoclinic form of hen egg-white lysozyme crystallized at 313°K — Comparison of 2 independent molecules." *Acta Crystallographica* **D50**, 250-257.

Kobe, B. and Deisenhofer, J. (1995) "A structural basis of the interactions between leucine-rich repeats and protein ligands." *Nature* **374**, 183-186.

Ladner, J.E., Wladkowski, B.D., Svensson, L.A., Sjölin, L. and Gilliland, G.L. (1997) "X-ray structure of a ribonuclease A-uridine vanadate complex at 1.3 Å resolution." *Acta Crystallographica* **D53**, 290-301.

Laskowski, R.A., Macarthur, M.W., Moss, D.S. and Thornton, J.M. (1993) "ProCheck - A program to check the stereochemical quality of protein structures." *Journal of Applied Crystallography* **26**, 283-291.

Lovell, S.C., Word, J.M., Richardson, J.S. and Richardson, D.C. (2000) "The penultimate rotamer library." *Proteins: Structure, Function, and Genetics* **40**, 389-408.

Richardson, D.C. and Richardson, J.S. (1992) "The kinemage: A tool for scientific illustration." *Protein Science* **1**, 3-9.

Richardson, J.S. and Richardson, D.C. (2001) "Mage, Probe, and kinemages", in "Crystallography of biological macromolecules." (Eds, Rossmann, M.G. and Arnold, E.) *International Tables for Crystallography,* **Vol. F,** 727-730. Kluwer Academic Publishers, The Netherlands, Dordrecht.

RichardsonLabWebSite (2001) http://kinemage.biochem.duke.edu. Richardson, D.C.

Shi, H. and Moore, P.B. (2000) "The crystal structure of yeast phenylalanine tRNA at 1.93 A resolution: A classic structure revisited." *RNA* **6**, 1091-1105.

Vriend, G. (1990) "What If: A molecular modeling and drug design program." *Journal of Molecular Graphics* **8**, 52-56.

Wlodawer, A., Borkakoti, N., Moss, D.S. and Howlin, B. (1986) "Comparison of two independently refined models of ribonuclease-A." *Acta Crystallographica* **B42**, 379-387.

Wlodawer, A., Svensson, L.A., Sjölin, L. and Gilliland, G.L. (1988) "Structure of phosphate-free ribonuclease a refined at 1.26 Å." *Biochemistry* **27**, 2705-2717.

Word, J.M., Bateman, R.C. Jr., Presley, B.K., Lovell, S.C. and Richardson, D.C. (2000) "Exploring steric constraints on protein mutations using Mage/Probe." *Protein Science* **9**, 2251-2259.

Word, J.M., Lovell, S.C., LaBean, T.H., Taylor, H.C., Zalis, M.E., Presley, B.K., Richardson, J.S. and Richardson, D.C. (1999a) "Visualizing and quantifying molecular goodness-of-fit: Small-probe contact dots with explicit Hydrogens." *Journal of Molecular Biology* **285**, 1711-1733.

Word, J.M., Lovell, S.C., Richardson, J.S. and Richardson, D.C. (1999b) "Asparagine and glutamine: Using hydrogen atom contacts in the choice of side-chain amide orientation." *Journal of Molecular Biology* **285**, 1735-1747.

Authors' Addresses

Jane S. Richardson, Department of Biochemistry, Duke University, Durham, NC 27710-3711, USA. Email: dcrjsr@kinemage.biochem.duke.edu.

David C. Richardson, Department of Biochemistry, Duke University, Durham, NC 27710-3711, USA.

Chapter 7

Protein Classification: A Geometric Hashing Approach

Xiong Wang and Jason T. L. Wang

7.1 Introduction

Protein classification has been a very important research topic [Kihara *et al.*, 1998; Pasquier and Hamodrakas, 1999; Wang *et al.*, 1999]. Traditionally, proteins are classified according to their functions. However, recently, many approaches have been proposed to classify proteins according to their structures, including secondary structures and three dimensional structures. Many of these methods complemented the traditional approach. We introduce here an algorithm that discovers frequently occurring patterns in a set of proteins, represented by 3D graphs, and use these patterns to classify the proteins. Our approach is a variant of the geometric hashing technique.

Proteins are large molecules, comprising hundreds of amino acids (residues) [Pu *et al.*, 1992; Wang *et al.*, 1999]. In each residue C_α, C_β and N atoms form a backbone of the residue [Pennec and Ayache, 1994]. Following [Vaisman *et al.*, 1998] we represent each residue by the three atoms. Thus, if we consider a protein as a 3D graph, then each node of the graph is an atom. Each node has a label, which is the name of the atom and is not unique in the protein. We assign a unique number to identify a node in the protein, where the order of numbering is obtained from Protein Data Bank (PDB) [Bernstein et al., 1977; Abola et al., 1987], accessible at http://www.rcsb.org. We construct substructures from a given set of proteins and evaluate the number of occurrences of each substructure in the data set. Those substructures that occur frequently are considered useful patterns and are used for classification.

7.2 Constructing Substructures

We discuss three different ways for constructing substructures. The first method segments a protein into consecutive substructures. The second and third methods construct substructures from the surface structure of the protein.

Segmenting a Protein

We decompose each protein into consecutive substructures, each substructure containing 6 atoms. Two adjacent substructures overlap by sharing the two neighboring atoms on the boundary of the two substructures (see Figure 1). Thus, each substructure is a portion of the polypeptide chain backbone of a protein where the polypeptide chain is made up of residues linked together by peptide bonds. The peptide bonds have strong covalent bonding forces that make the polypeptide chain rigid. As a consequence, the substructures used by our algorithm are rigid.

The Intuitive Protein Surface

Significant studies have shown that the structure of a protein surface is more related to the function of the protein. For example, Chirgadze and Larionova [1999] found that sign-alternating charge clusters are a common feature of the surface of a globular protein and they play a general functional role as a local polar factor of the protein surface. Rosen et al. [1998] examined the reliability of surface comparisons in searching for active sites in proteins. They suggested that, the detection of a patch of surface on one protein that is similar to an active site in another may indicate similarities in enzymatic mechanisms in enzyme functions, and implicate a potential target for ligand/inhibitor design.

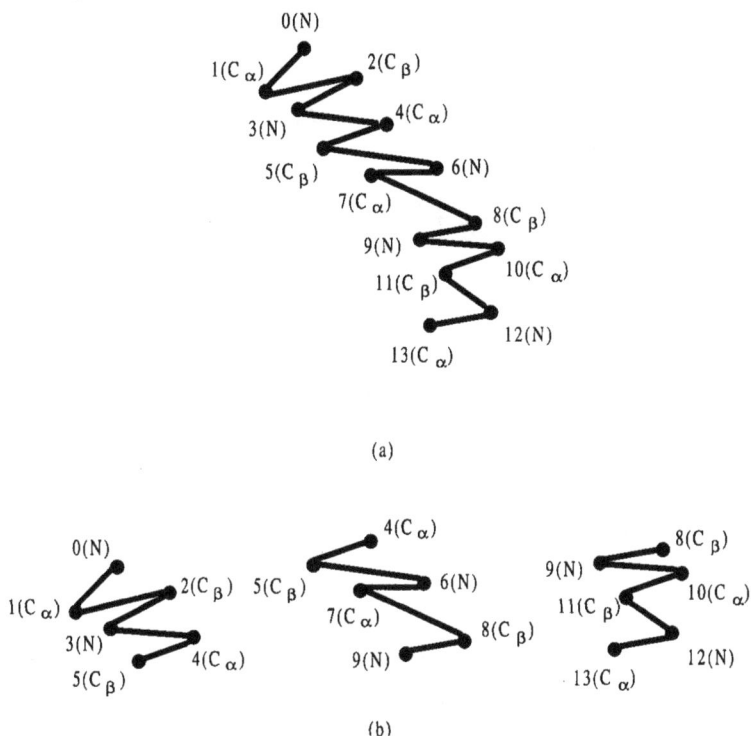

Figure 1. Segmenting a protein.

In our second method, we extract the surface atoms from a protein and construct substructures from the surface. Since each atom is represented by the coordinates of its center in a three dimensional space, for presentation purposes, we also use points to refer to these atoms. The idea here is to draw grids along each dimension and pick up the points with the maximal and minimal coordinates inside each strip. Figure 2 shows a two dimensional example. Inside each strip that is parallel to the Y-axis, the points with the maximal Y and minimal Y coordinates are surface points. Surface points are highlighted by solid balls in Figure 2. The lines connecting the surface points delineate the shape of the surface (see Figure 2).

Notice that, the width of the strips is a parameter, which can be adjusted.

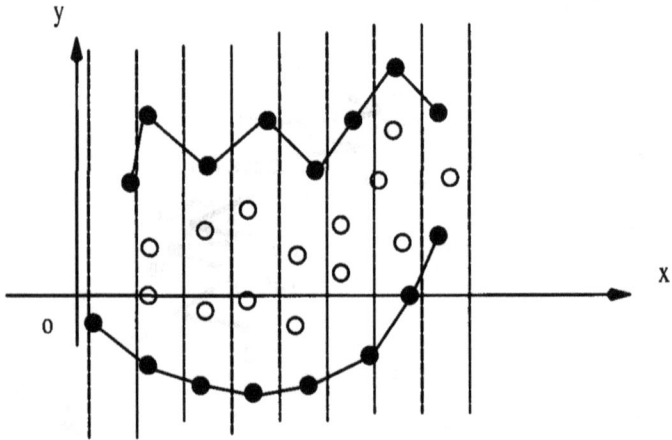

Figure 2. The surface of a protein.

For the same protein, the set of surface atoms can be different if this parameter is set to different values. Figure 3 shows a different set of surface atoms for the same protein in Figure 2.

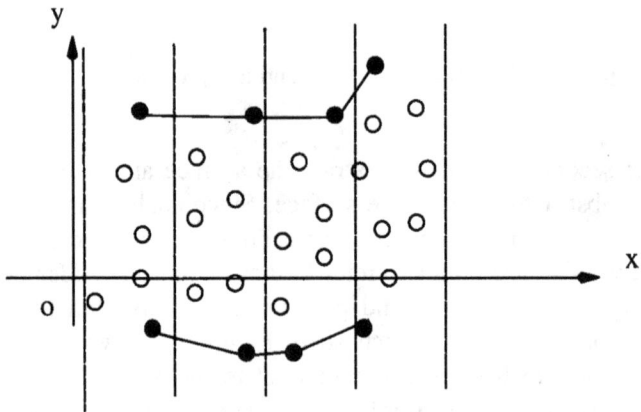

Figure 3. Another surface of the same protein.

Figure 4 shows the surface extracting algorithm along one dimension. Range is a parameter that can be adjusted. Notice that the algorithm in Figure 4 only finds the surface points along the Z axis. The same algorithm can be used to find surface points along the X and Y axes. These points are then combined together to form the surface of a protein.

Procedure Find_Surface
Input: A set of points D with coordinates.
Output: A set of points S that form part of the surface of D.

let X_{min} (X_{max}) be the minimal (maximal) X coordinates of all points in D, respectively;
let Y_{min} (Y_{max}) be the minimal (maximal) Y coordinates of all points in D, respectively;

for $(x = X_{min}; x < X_{max}; x = x + 2 * Range)$ **do**
    for $(y = Y_{min}; y < Y_{max}; y = y + 2 * Range)$ **do**
    begin
      in those points (x_p, y_p, z_p), such that
      $x_p \in (x - Range, x + Range]$ and
      $y_p \in (y - Range, y + Range]$,
      find the point p_{min} (p_{max}) whose Z coordinate is the smallest
        (largest), respectively;
      if $p_{min} \notin S$ **then**
        insert p_{min} into S;
      if $p_{max} \notin S$ **then**
        insert p_{max} into S;
    end

Figure 4. The surface extracting algorithm.

Let

$$R_x = X_{max} - X_{min}, \ R_y = Y_{max} - Y_{min}, \ R_z = Z_{max} - Z_{min}$$

and

$$C = \frac{R_x \times R_y + R_y \times R_z + R_x \times R_z}{2 \times Range}.$$

The complexity of the algorithm is $O\ (C \times |D|)$, where $|D|$ is the size of the protein.

Let D be the set of all the atoms in a protein and S be the surface atoms of the protein. For any atom $p \in S$, our second method considers p and its k-nearest neighbors in D as a substructure.

The α-Surface

Our third method employs a more formal definition of surface atoms, called α-surfaces. Our definition of α-surfaces is inspired by the definition of α-shapes, introduced by Edelsbrunner and Mücke [1994].

Definition 1 *Given a point O in the three dimensional Euclidean space R^3 and a real number $\alpha\ (0 < \alpha < \infty)$, an α-ball is the set of points $B(O,\alpha) = \{P \mid P \in R^3$ and $\| P\text{-}O \| < \alpha\}$, where $\| P\text{-}O \|$ is the Euclidean distance between P and O. A closed α-ball $\overline{B}(O,\alpha)$ is the α-ball $B(O,\alpha)$ plus its bounding sphere, i.e. $\overline{B}(O,\alpha) = \{P \mid P \in R^3$ and $\| P\text{-}O \| \le \alpha\}$.*

Definition 2 *Given a finite set D of discrete points in R^3 and a real number $\alpha\ (0 < \alpha < \infty)$, the α-surface S of D is defined as $S = \{ P \mid P \in D$ and $(\exists\ O \in R^3$ such that $B(O,\alpha) \cap D = \Phi$ and $P \in \overline{B}(O,\alpha))\}$. When $B(O,\alpha) \cap D = \Phi$ and $P \in \overline{B}(O,\alpha) \cap D$, we say that α-ball $B(O,\alpha)$ touches P. $P \in S$ is called a surface point with respect to α (or simply a surface point when the context is clear).*

Figures 5 and 6 show two α-surfaces of the same point set, with respect to two different α values.

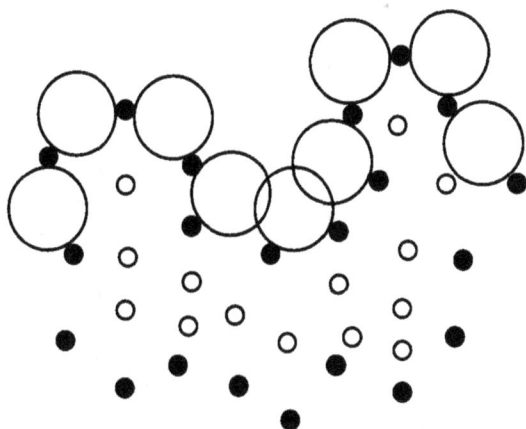

Figure 5. An α-surface in R^2.

The definition of α-surfaces is general. In the context of protein data, we need some adjustments. First of all, the surface of a protein is important to its function, because the protein reacts to its surroundings through its surface. Thus we are not concerned with those parts of α–surfaces that are not *visible*, namely those surface atoms that are enclosed inside the protein. Secondly, when α is small, the α-surface of D could be split to two pieces. A protein is one molecule. Its surface should be in one piece. We specify the adjustments in the following definition.

Definition 3 *Let α $(0 < \alpha < \infty)$ be a real number and S be the α-surface of a finite set D. S is connected, if for any two surface points P_1, $P_2 \in S$ there are a finite number of α-balls: $B(O_1,\alpha)$, $B(O_2,\alpha)$, ..., $B(O_n,\alpha)$, such that:*

(i) $B(O_i,\alpha) \cap D = \Phi$ $(1 \leq i \leq n)$.

(ii) $\overline{B}(O_i,\alpha) \cap \overline{B}(O_{i+1},\alpha) \cap S \neq \Phi$ $(1 \leq i \leq n-1)$.

(iii) $P_1 \in \overline{B}(O_1,\alpha)$.

(iv) $P_2 \in \overline{B}(O_n,\alpha)$.

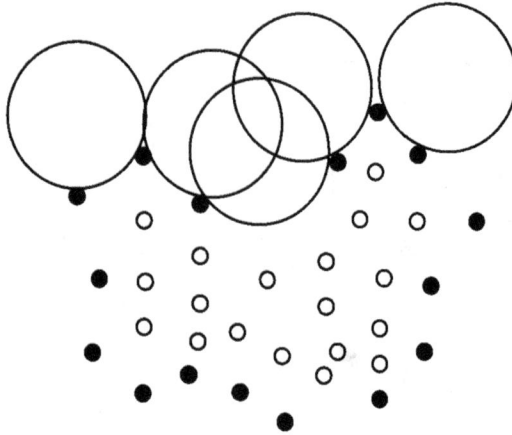

Figure 6. Anther α-surface w.r.t a different α value.

Notice that, (ii) requires two contiguous α-balls to touch on at least one common surface point. Imagine that the α-ball is solid, so are the points in D, and we roll the α-ball along the surface of D. Intuitively, if an α-surface is connected, we can roll an α-ball from one surface point to another along the surface.

Starting from the point with the maximum X-coordinate in D, the surface extracting algorithm rolls the α-ball to any surface point that can be touched in a breadth first manner.[3] The algorithm maintains a queue Q which holds a subset of the α-surface S that is under extension. The basic rolling procedure of the algorithm rolls the α-ball around one surface point in Q, so that all its neighboring points in S will be touched at least once by the α-ball. These neighbors are added to Q. Figure 7 illustrates the procedure. The α-ball is rolled around P_0 so that P_0's neighbors P_1, P_2, P_3, P_4, P_5 and P_6 are touched by the α-ball.

Since the neighboring surface points are within distance 2α of the current surface point, to speed up the process, we partition D at the very beginning. Let x_{min} (x_{max}) be the minimum (maximum) X coordinate of all the points in D, respectively. Let x_0, x_1, ..., x_n be defined as follows:

---

[3] Obviously, the point with the maximum X coordinate in D is a surface point with respect to any α.

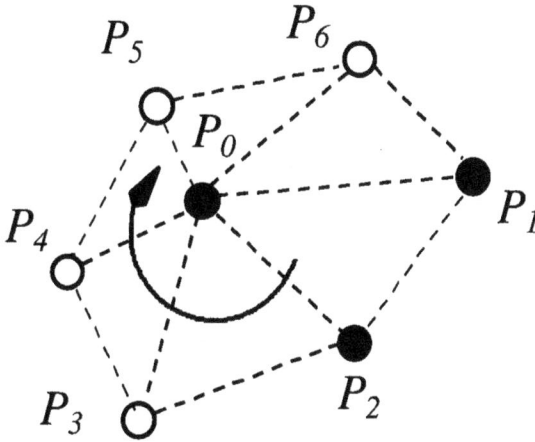

Figure 7. Rolling an α-ball.

(i) $x_0 = x_{min}$;

(ii) $x_{i+1} = x_i + 2\alpha$ $(0 \le i \le n\text{-}1)$; and

(iii) $x_{n-1} \le x_{max}$ and $x_{max} < x_n$.

We cut the range $[x_{min}, x_{max}]$ to segments $[x_i, x_{i+1}]$ $(0 \le i \le n\text{-}1)$ with length 2α. Similarly, let y_{min} (y_{max}) be the minimum (maximum) Y coordinate and z_{min} (z_{max}) be the minimum (maximum) Z coordinate, respectively. We cut the ranges $[y_{min}, y_{max}]$ and $[z_{min}, z_{max}]$ to segments with length 2α. Each partition $Pt_{i,j,k}$ is a cube $Pt_{i,j,k} = \{(x, y, z) \mid x_i \le x < x_{i+1}, y_j \le y < y_{j+1}, \text{ and } z_k \le z < z_{k+1}\}$. Figure 8 shows a two dimensional example.

For any given point $P = (x, y, z) \in D$, let

$$i = \left\lceil \frac{x - x_{min}}{2\alpha} \right\rceil, \quad j = \left\lceil \frac{y - y_{min}}{2\alpha} \right\rceil \text{ and } k = \left\lceil \frac{z - z_{min}}{2\alpha} \right\rceil$$

P belongs to partition $Pt_{i,j,k}$ and the points that are within distance 2α of P are all located in the partitions surrounding $Pt_{i,j,k}$.

Assuming that the points in D are evenly distributed, the complexity of the surface extracting algorithm is:

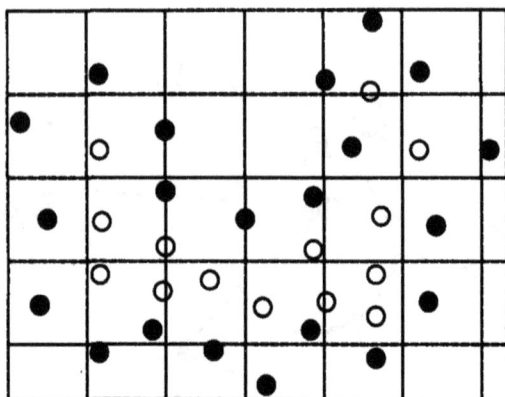

Figure 8. Partitioning points in a two dimensional space.

$$O\left(\frac{|D|^2}{N}\right),$$

where $|D|$ is the size of D and

$$N = \left[\frac{X_{max} - X_{min}}{2\alpha}\right] \times \left[\frac{Y_{max} - Y_{min}}{2\alpha}\right] \times \left[\frac{Z_{max} - Z_{min}}{2\alpha}\right]$$

is the total number of partitions.

Let S be the α-surface of a protein. For any atom $p \in S$, our third method considers p and its k-nearest neighbors in S as a substructure. Notice the difference between this third method and the second method depicted earlier.

7.3 Discovering Frequently Occurring Patterns

Given a set of substructures of proteins, this section presents the algorithm to evaluate their frequency of occurrence. The algorithm proceeds in two phases: the hashing phase and the evaluation phase. In the hashing phase, given a substructure *Str* of a protein, we attach a local coordinate frame, called a Substructure Frame *SF*, to a node P_0 in *Str* (see Figure 9).

Suppose the coordinates of P_0 are (x_0, y_0, z_0). This local coordinate frame is represented by three basis points P_{b1}, P_{b2} and P_{b3}, with coordinates P_{b1} (x_0, y_0, z_0), P_{b2} (x_0+1, y_0, z_0), and P_{b3} (x_0, y_0+1, z_0), respectively. The origin is P_{b1} and the three basis vectors are $V_{b1,b2}$, $V_{b1,b3}$ and $V_{b1,b2} \times V_{b1,b3}$. Here, $V_{b1,b2}$ represents the vector starting at point P_{b1} and ending at point P_{b2}. $V_{b1,b2} \times V_{b1,b3}$ stands for the cross product of the two corresponding vectors. We hash all three-node combinations, referred to as node-triplets, in the substructure *Str* to a 3D hash table. Notice that, three sorted nodes uniquely determine another coordinate frame (see Figure 10).

Let P_i, P_j and P_k be three nodes, such that $\| P_i - P_j \| \leq \| P_i - P_k \| \leq \| P_j - P_k \|$, where $\| P_i - P_j \|$, $\| P_i - P_k \|$, and $\| P_j - P_k \|$ stand for the Euclidean distance between each pair of the nodes. The local coordinate frame, denoted *LF[i, j, k]*, is constructed, using $V_{i,j}$, $V_{i,k}$ and $V_{i,j} \times V_{i,k}$ as basis vectors. With respect to *LF[i, j, k]*, the coordinates of P_{b1}, P_{b2}, P_{b3} are geometric invariants. We store these coordinates together with a protein identification number and a substructure number in the hash table. The hash bin addresses are calculated using $\| P_i - P_j \|^2$, $\| P_i - P_k \|^2$ and $\| P_j - P_k \|^2$. At the end of the hashing phase, all substructures are stored in the hash table.

In the evaluation phase, the algorithm considers each substructure as a candidate pattern and rehashes it to evaluate its number of occurrences in the training data. In this phase, we again take each node-triplet from the candidate pattern and utilize the lengths of the three sorted edges to access the hash table. All the triplets that were stored in the accessed hash bin are recognized as matches and their local coordinate systems *SFs* are recovered based on the global coordinate system in which the candidate pattern is given. The triplet matches are augmented to larger substructure matches when they come from the same substructure and their recovered local coordinate systems match each other (see Figure 11).

A candidate pattern *M* occurs in the protein or on the surface of a protein if *M* matches any substructure from the protein within one mutation. A candidate pattern *M* matches a substructure *Str* with *n* mutations if by applying an arbitrary number of rotations and translations as well as *n* node insert/delete operations to *M*, one can transform *M* to *Str* (see [Wang *et al.*, 1997; Wang *et al.*, 2002] for details).

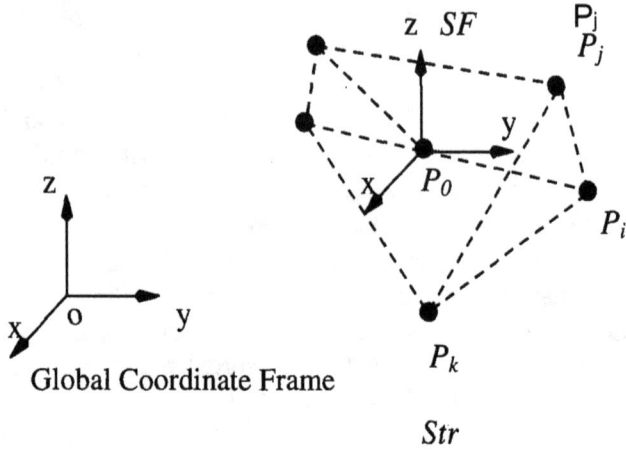

Figure 9. A substructure and the SF attached to the substructure.

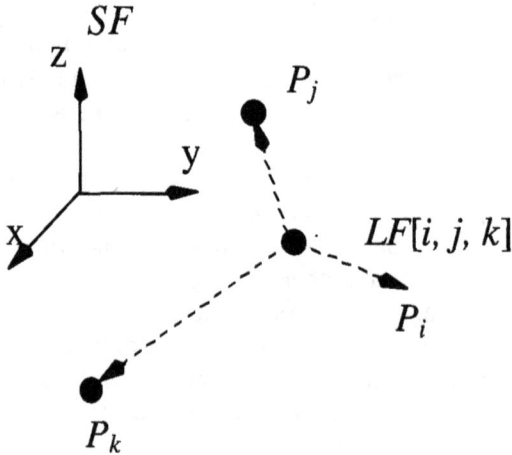

Figure 10. A node-triplet and the local coordinate frame $LF[i, j, k]$.

Two Triplets from the Hash Table

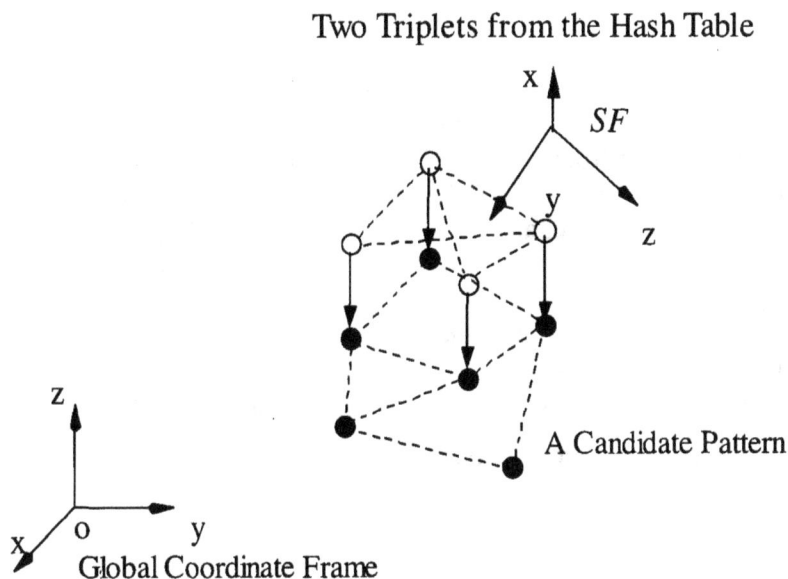

Figure 11. Two triplet matches that are augmentable.

7.4 Classifying Proteins

We applied our techniques to classifying three families of proteins. Since the α-surface technique achieves the best result, we report that result only. For each family i of the proteins, we identify two types of patterns on the surfaces of the training data, the *pro* patterns and the *con* patterns. The pro patterns occur more frequently in family i than in the other two families. The con patterns occur more frequently in the other two families than in family i. Each candidate pattern M found on the surfaces of the training data is associated with two weights pro^i and con^i where

$$pro^i = \frac{n_i - \max_{j \in \{1,2,3\} - \{i\}}\{n_j\}}{\max_{j \in \{1,2,3\} - \{i\}}\{n_j\} + 1}$$

$$con^i = \frac{\min_{j \in \{1,2,3\} - \{i\}} \{n_j\} - n_i}{n_i + 1}$$

Here n_i is M's occurrence number in the training data of family i. We add denominators to both weights because we observed that some patterns are common to proteins from different families. Although they may still occur more frequently in some family, they really are not specific to any family. For each family we collect all the patterns having a weight greater than zero and use them as pro patterns and con patterns of that family, respectively. It can be proved that any pattern M that occurs in the training data is either a pro pattern or a con pattern of some family, unless M's occurrence numbers tie in all the three families.

We classify a test protein Q in the following way. Let $M^i_1, \ldots, M^i_{pi}$ be all the pro patterns for family i. Family i obtains a pro score

$$N^i_{pro} = \frac{\sum_{k=1}^{pi} d_k \times pro^i_k}{\sum_{k=1}^{pi} pro^i_k}$$

where

$$d_k = \begin{cases} 1 & \text{if } M^i_k \text{ occurs in } Q \\ \\ 0 & \text{otherwise} \end{cases}$$

and pro^i_k is the weight associated with M^i_k. The protein Q is classified to the family i with maximum N^i_{pro}. We add the denominator to make the score fair to all families. Notice that the maximum possible score for any family is 1. If we can not decide a winner from the pro scores, e.g. the scores are the same for two families, the con patterns are used to break the tie. Let $T^i_1, \ldots,$ T^i_{qi} be all the con patterns for family i. Family i obtains a con score

$$N^i_{con} = \frac{\sum_{k=1}^{qi} d_k \times con^i_k}{\sum_{k=1}^{qi} con^i_k}$$

where

$$d_k = \begin{cases} 1 & \text{if } T_k^i \text{ occurs in } Q \\ \\ 0 & \text{otherwise} \end{cases}$$

and con_k^i is the weight associated with T_k^i. The protein Q is classified to the family i with minimum N_{con}^i. If we still can not decide a winner, then the "no-opinion" verdict is given.

7.5 Experimental Results

We have implemented all algorithms using C++ on a Sun Ultra 10 workstation running Solaris 8. We selected three families of proteins from SCOP [Murzin *et al.*, 1995]. SCOP is accessible at http://scop.mrc-lmb.cam.ac.uk/scop/. The three families pertain to Transmembrane Helical Fragments, Matrix Metalloproteases -- catalytic domain, and Immunoglobulin -- I set domains. In determining the structure of a protein, we consider only C_α, C_β and N atoms. Figure 12 shows a protein whose PDB Code is 1cqr. It has 1089 atoms in the backbone. Figure 13 shows an α-surface found by the proposed algorithm, with respect to α=7.5. It has 242 atoms.

We classified the proteins as described in Section 7.4. When adjusting α in the surface extracting algorithm, we found that $\alpha = 7.5$ yielded the best result. When constructing substructures (patterns), we found the substructures with 6 points yielded the best result. In each of these substructures, there was a surface point together with its 5 nearest neighbors on the α-surface. The algorithm produced a set of surface points from a protein that were on average 25% of the size of the protein.

We use recall (R) and precision (P) to evaluate the effectiveness of our classification algorithm. Recall is defined as

$$R = \frac{T - \sum_{i=1}^{3} M^i}{T} \times 100\%$$

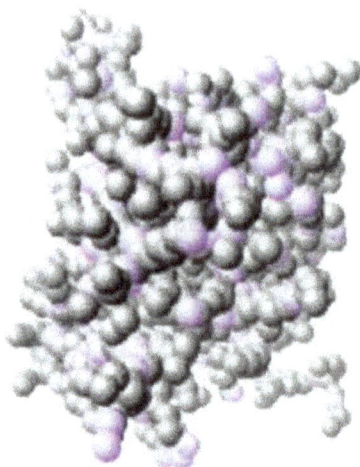

Figure 12. A protein (1cqr).

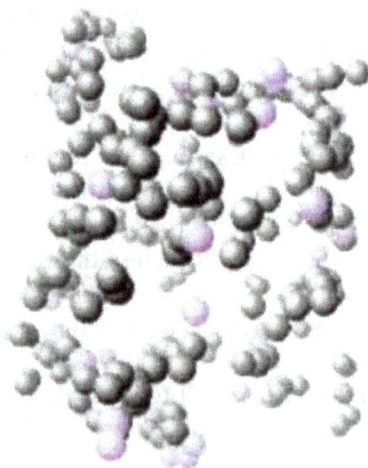

Figure 13. An α-surface of the protein (1cqr) in Figure 12.

where T is the total number of test proteins and M^i is the number of test proteins that belong to family i but are not assigned to family i by our algorithm (they are either assigned to family j, $j \neq i$, or they receive the "no-opinion" verdict). Precision is defined as

$$P = \frac{T - \sum_{i=1}^{3} G^i}{T} \times 100\%$$

where G^i is the number of test proteins that do not belong to family i but are assigned by our algorithm to family i. With the 10-way cross validation scheme,[4] the average R over the ten tests was 93.7% and the average P was 95.2%. It was found that 4.3% test proteins on average received the "no-opinion" verdict during the classification.

7.6 Conclusion

We investigate approaches to the discovery of frequently occurring patterns in three dimensional structures and their application to protein classification. Some ideas described here have appeared in [Wang, 2001a; Wang, 2001b; Wang and Wang, 2001; Wang *et al.*, 2002]. Future work includes extending our algorithms to build a structure-based search engine for proteins.

References

Abola, E.E., Bernstein, F.C., Bryant, S.H., Koetzle, T.F. and Weng, J. (1987) "Protein Data Bank." In Allen, F.H., Bergerhoff, G. and Sievers, R. editors, *Data Commission of the International Union of Crystallography*, 107-132, Bonn/Cambridge/Chester.

Bernstein, F.C., Koetzle, T.F., Williams, G.J.B., Meyer, E.F., Brice, M.D., Rodgers, J.R., Kennard, O., Shimanouchi, T. and Tasumi, M. (1977) "The Protein Data

---

[4] That is, each family was divided into 10 groups of roughly equal size. Ten tests were conducted. In each test, a group was taken from a family and used as test data; the other nine groups were used as training data for that family.

Bank: A computer-based archival file for macromolecular structures." *Journal of Molecular Biology* **112**, 535-542.

Chirgadze, Y.N. and Larionova, E.A. (1999) "Spatial sign-alternating charge clusters in globular proteins." *Protein Engineering* **12**, 101-105.

Edelsbrunner, H. and Mücke, E.P. (1994) "Three-dimensional alpha shapes." *ACM Transactions on Graphics* **13(1)**, 43-72.

Kihara, D., Shimizu, T. and Kanehisa, M. (1998) "Prediction of membrane proteins based on classification of transmembrane segments." *Protein Engineering* **11**, 961-970.

Murzin, A.G., Brenner, S.E., Hubbard, T. and Chothia, C. (1995) "SCOP: A structural classification of proteins database for the investigation of sequences and structures." *J. Mol. Biol.* **247**, 536-540.

Pasquier, C. and Hamodrakas, S.J. (1999) "A hierarchical artificial neural network system for the classification of transmembrane proteins." *Protein Engineering* **12**, 631-634.

Pennec, X. and Ayache, N. (1994) "An $O(n^2)$ algorithm for 3D substructure matching of proteins." *Proceedings of the 1st International Workshop on Shape and Pattern Matching in Computational Biology*, 25-40.

Pu, C., Sheka, K.P., Chang, L., Ong, J., Chang, A., Alessio, E., Shindyalov, I.N., Chang, W. and Bourne, P.E. (1992) "PDBtool: A prototype object oriented toolkit for protein structure verification." Technical Report CUCS-048-92, Department of Computer Science, Columbia University.

Rosen, M., Lin, S.L., Wolfson, H. and Nussinov, R. (1998) "Molecular shape comparisons in searches for active sites and functional similarity." *Protein Engineering* **11**, 263-277.

Vaisman, I.I., Tropsha, A. and Zheng, W. (1998) "Compositional preferences in quadruplets of nearest neighbor residues in protein structures: Statistical geometry analysis." *Proceedings of the IEEE International Join Symposia on Intelligence and Systems*, 163-168.

Wang, J.T.L., Shapiro, B.A. and Shasha, D. (1999) *Pattern Discovery in Biomolecular Data: Tools, Techniques and Applications*. Oxford University Press, New York.

Wang, X. (2001a) "Mining protein surfaces." 2001 ACM SIGMOD Workshop on Research Issues in Data Mining and Knowledge Discovery, Santa Barbara, California.

Wang, X. (2001b) "α-Surface and its application to mining protein data." The First IEEE International Conference on Data Mining, San Jose, California.

Wang, X. and Wang, J.T.L. (2001) "Analyzing protein surface for classification: A geometric hashing approach." *Proceedings of the Atlantic Symposium on Computational Biology, Genome Information Systems & Technology*, 31-34.

Wang, X., Wang, J.T.L., Shasha, D., Shapiro, B.A., Dikshitulu, S., Rigoutsos, I. and Zhang, K. (1997) "Automated discovery of active motifs in three dimensional molecules." *Proceedings of the 3rd International Conference on Knowledge Discovery and Data Mining*, 89-95.

Wang, X., Wang, J.T.L., Shasha, D., Shapiro, B.A., Rigoutsos, I. and Zhang, K. (2002) "Finding patterns in three dimensional graphs: Algorithms and applications to scientific data mining." *IEEE Transactions on Knowledge and Data Engineering* **14(4)**, 731-749.

Authors' Addresses

Xiong Wang, Department of Computer Science, California State University, Fullerton, CA 92834, USA. Email: wang@ecs.fullerton.edu.

Jason T.L. Wang, Department of Computer Science, College of Computing Sciences, New Jersey Institute of Technology, University Heights, Newark, NJ 07102, USA. Email: wangj@oak.njit.edu.

Chapter 8

Interrelated Clustering: An Approach for Gene Expression Data Analysis

Chun Tang, Li Zhang, Aidong Zhang and Murali Ramanathan

8.1 Introduction

Array technologies are capable of simultaneously measuring the signals for thousands of messenger RNAs and large numbers of proteins from single samples. Arrays are now widely used in basic biomedical research for mRNA expression profiling and are increasingly being used to explore patterns of gene expression in clinical research [Schena *et al.*, 1995; DeRisi *et al.*, 1996; Schena *et al.*, 1996; Shalon *et al.*, 1996; Heller *et al.*, 1997; Chen *et al.*, 1998; Ermolaeva *et al.*, 1998; Welford *et al.*, 1998; Iyer *et al.*, 1999]. The customary approach in array analysis is to obtain data from fluorescence scanners or phosphorimagers and to analyze the array images using dedicated image analysis software, usually provided by the array manufacturer. Minimally, these software identify spots and analyze spot intensities, map spots to genes, and condition of the data. The normalized results are exported as flat tables to other software where a typical preliminary analysis may involve exploratory cluster analysis, biostatistical analysis and bioinformatics research for interesting genes.

The raw microarray data are images which can then be transformed into gene expression matrices where usually the rows represent genes, and the columns represent various samples. The numeric value in each cell characterizes the expression level of the particular gene in a particular sample. Innovative techniques to efficiently and effectively analyze these fast growing gene data are required, which will have a significant impact on the field of bioinformatics. But the high-dimensionality and size of array-derived data poses challenging problems in both computational and biomedical research, and the difficult task ahead is converting genomic data into knowledge. Various methods have been developed using both traditional and innovative techniques to extract, analyze, and visualize gene expression data generated from DNA microarrays.

The existing data-clustering methods fall into two major categories: supervised clustering and unsupervised clustering. The supervised approach assumes that additional information is attached to some (or all) data; for example, samples are labeled as diseased vs. normal. Using this information, a classifier can be constructed to predict the labels from the expression profile. The major supervised clustering methods include neighborhood analysis [Golub *et al.*, 1999], the support vector machine [Brown *et al.*, 2000; Furey *et al.*, 2000; Pavlidis *et al.*, 2001], the tree harvesting method [Hastie *et al.*, 2001], the decision tree method [Zhang *et al.*, 2001], statistical approaches such as the maximum-entropy model [Jiang *et al.*, 2001], and a variety of ranking-based methods [Moler *et al.*, 2000; Park *et al.*, 2001; Thomas *et al.*, 2001].

Unsupervised approaches assume little or no prior knowledge. The goal of such approaches is to partition the set of samples or genes into statistically meaningful classes [Ben-Dor *et al.*, 2001]. A typical example of unsupervised data analysis is to find groups of co-regulated genes or related samples. Currently most of the research focuses on the supervised analysis; relatively less attention has been paid to unsupervised approaches in gene expression data analysis which is important in a context where little domain knowledge is available [Spellman *et al.*, 1998; Barash and Friedman, 2001]. The hierarchical clustering method [Eisen *et al.*, 1998; Alizadeh *et al.*, 2000; Zou *et al.*, 2000; Hakak *et al.*, 2001; Herrero *et al.*, 2001; Martin *et al.*, 2001; Virtaneva *et al.*, 2001; Welsh *et al.*, 2001], the k-means clustering algorithms [Hartigan, 1975; Tavazoie *et al.*, 1999; Han and Kamber, 2000] and the self-organizing maps [Kohonen, 1984; Golub *et al.*, 1999; Tamayo *et al.*, 1999; Holter *et al.*, 2000; Mody *et al.*, 2001] are the major unsupervised clustering methods which have been commonly applied to various data sets.

Information in gene expression data can be studied in two angles [Brazma and Vilo, 2000]: analyzing expression profiles of genes by comparing rows in the expression matrix [Eisen *et al.*, 1998; Ben-Dor *et al.*, 1999; Perou *et al.*, 1999; Tamayo *et al.*, 1999; Alter *et al.*, 2000; Brown *et al.*, 2000; Manduchi *et al.*, 2000; Hastie *et al.*, 2001] and analyzing expression profiles of samples by comparing columns in the matrix [Golub *et al.*, 1999; Azuaje, 2000; Slonim *et al.*, 2000]. While most researchers focus on either genes or samples, in a few occasions, sample clustering has been combined with gene clustering. Alon *et al.* [1999] proposed a partitioning-based algorithm to study 6500 genes of 40 tumor and 22 normal colon tissues for clustering genes and samples independently. Getz *et al.* [2000] proposed a coupled two-way clustering method to identify subsets of both genes and samples. Xing and Karp [2001] proposed a clustering method called CLIFF which iteratively uses sample partitions as a reference to filter genes. None of these approaches offers a definitive solution to the fundamental challenge of detecting meaningful patterns in the samples while pruning out irrelevant genes in a context where little domain knowledge is available.

In this chapter, we will introduce an interrelated two-way clustering approach for unsupervised analysis of gene expression data. Unlike previous work mentioned above, in which genes and samples were clustered either independently or both data being reduced, our approach is to delineate the relationships between gene clusters and sample partitions while conducting an iterative search for sample patterns and detecting significant genes of empirical interest. The performance of the proposed method will be illustrated in the context of various data sets.

The remainder of this chapter is organized as follows. Section 8.2 introduces the interrelated clustering approach. Section 8.3 presents the experimental results on the multiple sclerosis data set and other data sets. And finally, the conclusion is provided in Section 8.4.

8.2 Interrelated Clustering

8.2.1 Motivation

Gene expression data can be viewed as matrices where rows represent genes, and columns represent samples such as tissues or experimental conditions. Let $G=\{g_1,...,g_i,...,g_n\}$ be the set of all genes, $S=\{s_1,...,s_j,...,s_m\}$ be the set of all samples, and $w_{i,j}$ be the intensity value associated with each gene g_i and sample s_j in the matrix. Thus the gene expression matrix $W=\{\ w_{i,j} \mid 1 \le i \le n,\ 1 \le j \le m\ \}$ has n rows (genes) and m columns (samples). Clustering can be used to group genes that manifest similar expression patterns for a set of samples [Eisen *et al.*, 1998; Ben-Dor *et al.*, 1999; Alter *et al.*, 2000; Brown *et al.*, 2000; Hastie *et al.*, 2001]. This view considers the $N=g_n$ genes as objects to be clustered, each represented by its expression profile, as a point in a $D=s_m$ dimensional space, measured over all of the samples. Another type of clustering is to cluster samples into homogeneous groups which may correspond to particular macroscopic phenotypes, such as clinical syndromes or cancer types [Golub *et al.*, 1999; Azuaje, 2000; Slonim *et al.*, 2000]. In this instance, the $N=s_m$ samples are viewed as the objects to be clustered, with the levels of expression of g_n genes playing the role of the features, representing each sample as a point in a $D=g_n$ dimensional space.

Sample clustering presents interesting but also very challenging problems. In typical microarray data sets, the sample space and gene space are of very different dimensionality, for example, $10^1 \sim 10^2$ samples versus $10^3 \sim 10^4$ genes. Clustering on the original high dimensional data is not guaranteed to capture a meaningful partition corresponding to empirical interest because [Xing and Karp, 2001]:

(1) A gene expression matrix is usually generated according to some actual empirical interest, like diseased vs. healthy condition for samples. But the same set of samples may also display gender, age, or other variability.

(2) Microarrays are not typically task-specific and most of the genes are not necessarily of interest. Sample-pattern detection is subject to interference from the large number of irrelevant or redundant genes which should be pruned out or filtered when clustering samples.

(3) For unsupervised analysis, uncertainty about which genes are relevant makes it difficult to construct an informative gene space to detect real sample partition.

To address these problems, we propose a framework for the unsupervised gene expression data analysis. In this framework, a pre-processing procedure is first applied to identify the relative important genes. Then an interrelated two-way clustering approach is applied to the gene expression matrix W. The goal of the interrelated two-way clustering involves two tasks: detection of meaningful patterns within the samples and selection of those significant genes which contribute to the samples' empirical pattern. To be more specific, they are:

(1) To select a subset of genes, usually called important genes, which are highly associated with the samples' experimental distributions. This can also be considered as genes filtering.

(2) To cluster the samples into different groups. According to the most popular experimental platforms, the number of different groups is usually two, for example, diseased samples and health control samples.

These two tasks are actually interconnected. Once the important genes are identified, the dimensions of the data will be efficiently reduced so to allow conventional clustering algorithms to be used to cluster samples. Conversely, once the salient sample patterns have been found, genes can be sorted for importance using similarity scores, such as correlation coefficient with the pattern. In general, if either an accurate sample partition or a set of significant genes is known, the other can then be easily obtained by supervised approaches [Golub *et al.*, 1999; Jorgensen, 2000; Jiang *et al.*, 2001]. With unsupervised clustering, however, factors such as the sparsity of data, the high dimensionality of the gene space, and the high percentage of irrelevant or redundant genes make it very difficult either to classify samples using traditional clustering algorithms [Hartigan, 1975; Hartigan and Wong, 1979] or pick out substantial genes in a context where little domain knowledge is available.

Since the volume of genes is large and no information regarding the actual partition of the samples assumed to be available, we cannot directly identify the sample patterns or significant genes. Rather, these goals must be gradually approached. First, we use the relationships of sample clusters and gene groups thus discovered to post a partial or approximate pattern. We then use this pattern to direct the elimination of irrelevant genes. In turn, the remaining meaningful genes will guide further sample pattern detection. Thus, we can formulate the problem of pattern discovery in the original data via an interplay between approximate partition detection and irrelevant gene pruning. Because of the complexity of the matrix, this procedure usually requires several iterations to achieve satisfactory results.

8.2.2 Pre-processing

In the gene expression matrix, different genes have different ranges of intensity values. Thus, the absolute intensity values of genes alone may be difficult to be compared, and thus may not indicate any significant relationships among the genes. But the relative values are more intrinsic. So we first normalize the original gene intensity values into relative values [Schuchhardt *et al.*, 2000; Yang *et al.*, 2001].

Our general formula for normalization is

$$w'_{i,j} = \frac{w_{i,j} - \mu_i}{\mu_i}, \text{ where } \mu_i = \frac{\sum_{j=1}^{m} w_{i,j}}{m} \tag{1}$$

In Formula 1, $w'_{i,j}$ denotes normalized intensity value for gene i of sample j, $w_{i,j}$ represents the original intensity value for gene i of sample j, m is the number of samples, and u_i is the mean of the intensity values for gene i over all samples.

As we mentioned earlier, among thousands of genes, not all of them have the same contribution in distinguishing the classes. Actually, some genes have little contribution. We need to remove those genes which have little reaction to the experiment condition. We believe that genes whose intensity values keep invariant or change very little among samples should be the ones to be removed. Figure 1 shows an example of gene distributions. In the figure, the horizontal axis represents samples. Each polygonal line indicates that a gene changing level varies among samples. The red-solid lines represent the genes whose relative intensity values vary little through all samples, and the blue-dash lines represent the genes whose intensity values vary much among samples.

Let's assume we have n genes and m samples. We denote each gene vector (after normalization) as

$$g_i = (w'_{i,1}, w'_{i,2}, ..., w'_{i,m}), \tag{2}$$

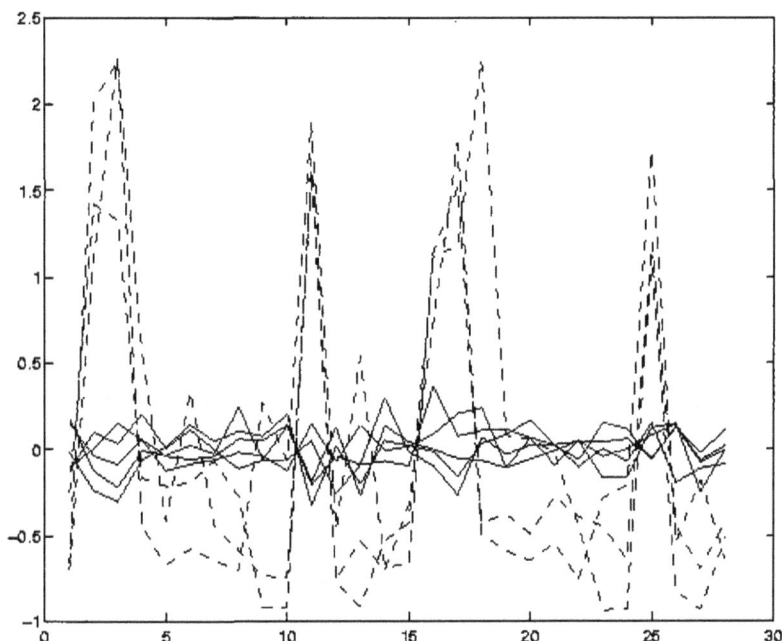

Figure 1. Genes intensity value distributions after normalization.

where $i = 1,2,...n$ for genes. We use *vector-cosine* between each gene vector and a pre-defined stable pattern E to test whether a normalized gene intensity value varies much among samples. The pattern is denoted as E= ($e_1, e_2, ...,$ e_m), where all e_i are equal.

$$\cos(\theta) = \frac{< \vec{g}_i, \vec{E} >}{\| \vec{g}_i \| \cdot \| \vec{E} \|} = \frac{\sum_{j=1}^{m} w'_{i,j} \times e_j}{\sqrt{\sum_{j=1}^{m} w'^2_{i,j}} \times \sqrt{\sum_{j=1}^{m} e_j^2}}, \qquad (3)$$

where θ is the *angle* between two vectors $\vec{g}_i$ and $\vec{E}$ in m-dimensional space. If the two vector patterns are highly similar, the vector-cosine will be close to *1*. The extreme cases are when two vectors are parallel, the vector-cosine value is *1*, but the vector-cosine value of two perpendicular vectors is *0*. After calculating vector-cosine values, we can choose a threshold to remove those genes which match pattern E closely (that is, those genes whose vector-cosine values with E are higher than the threshold, which means these genes change little during the experiment). Usually we can remove twenty to thirty percent of genes by this step, thus facilitating clustering in the next stage.

8.2.3 Interrelated Clustering

To perform interrelated two-way clustering, a distance measure to be used during the clustering procedure should be carefully chosen. One commonly used distance is the *Euclidean* distance. But for gene data, the similarity of the patterns between gene vectors seems more important than their spatial distance [Golub *et al.*, 1999; Jorgensen, 2000]. So we choose *correlation coefficient* [Devore, 1991], which measures the strength of linear relationship between two gene vectors. This measure has the advantage of calculating similarity depending only on the patterns between gene vectors but not on the absolute magnitude of the spatial vector. The formula of correlation coefficient between two vectors $X=(x_1,x_2,...x_k)$ and $Y=(y_1,y_2,...y_k)$ is defined as:

$$\rho_{x,y} = \frac{k(\sum_{i=1}^{k} x_i \times y_i) - (\sum_{i=1}^{k} x_i) \times (\sum_{i=1}^{k} y_i)}{\sqrt{\left[\sum_{i=1}^{k} x_i^2 - (\sum_{i=1}^{k} x_i)^2\right]\left[k\sum_{i=1}^{k} y_i^2 - (\sum_{i=1}^{k} y_i)^2\right]}} \tag{4}$$

where k is the length of vectors X and Y.

In interrelated two-way clustering, both genes and samples are simultaneously clustered. Dynamic relationship between gene clustering and sample clustering is used to reduce the number of genes into a reasonable size and perform class discovery. Our algorithm, illustrated in Figure 2, is an iterative procedure based on G with n_1 genes after pre-processing. The idea is to dynamically use the relationships between the groups of the genes and samples while iteratively clustering through both genes and samples to extract important genes and classify samples simultaneously. Within each iteration there are five main steps:

Step 1: clustering on genes. The task of this step is to cluster n_1 genes into k groups, denoted as G_i ($1 \le i \le k$), each of which is an exclusive subset of G. The clustering method can be any method for which we can give the

```
┌─────────────────────────────┐
│       pre-processing        │
└─────────────────────────────┘
                │
                ▼ ◄──────────────────────────┐
  ╭─────────────────────────────────╮        │
  │      clustering on genes        │        │
  ╰─────────────────────────────────╯        │
                │                             │
 *using each gene group*                      │
                ▼                             │
  ╭─────────────────────────────────╮        │
  │     clustering on samples       │        │
  ╰─────────────────────────────────╯        │
                │                             │
 *using two cluster results*                  │
                ▼                             │
  ╭─────────────────────────────────╮        │
  │  clustering results combination │        │
  ╰─────────────────────────────────╯        │
                │                             │
                ▼                             │
  ╭─────────────────────────────────╮        │
  │   finding heterogeneous groups  │        │
  ╰─────────────────────────────────╯        │
                │                             │
 *define patterns*                            │
                ▼                             │
  ╭─────────────────────────────────╮        │
  │      sorting and reducing       │        │
  ╰─────────────────────────────────╯        │
                │                             │
 *cross-validation*                           │
                ▼                             │
        ◄───────────────────►────────────────┘
          termination condition
              (diamond)
                │
                ▼
```

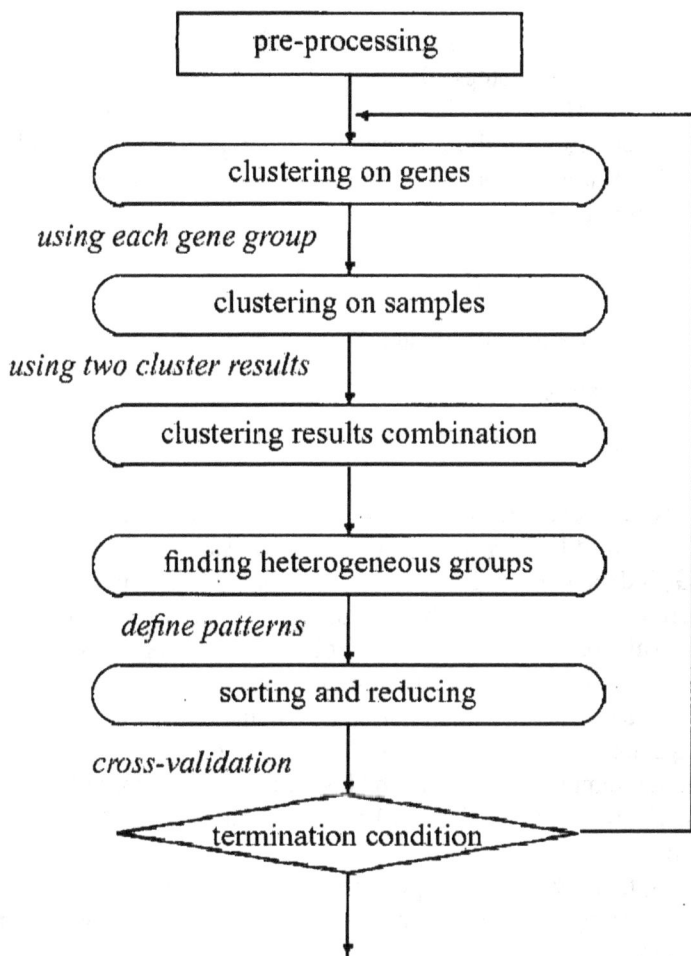

Figure 2. The structure of interrelated two-way clustering.

number of clusters, such as K-means or SOM [Hartigan, 1975; Hartigan and Wong, 1979].

Step 2: clustering on samples. Based on each group G_i *($1 \leq i \leq k$),* we independently cluster samples into two clusters (according to the most popular experimental conditions [Brazma and Vilo, 2000]), represented by $S_{i,a}$ and $S_{i,b}$.

Step 3: clustering results combination. This step combines the clustering results of the previous steps. By *Step 2*, we get k pairs of samples clusters $S_{i,a}$, $S_{i,b}$ $(i=1,2..k)$. Then we choose one cluster from each pair and find all possible intersection of these k sample clusters, denoted as sample groups C_j $(1 \leq j \leq 2^k)$. Without loss of the generality, let $k=2$. Then the samples can be divided into four groups:

- C_1 (intersection of $S_{1,a}$ and $S_{2,a}$);

- C_2 (intersection of $S_{1,a}$ and $S_{2,b}$);

- C_3 (intersection of $S_{1,b}$ and $S_{2,a}$);

- C_4 (intersection of $S_{1,b}$ and $S_{2,b}$).

Figure 3 illustrates the results of this combination. In the figure, the second and third lines show cluster results on samples based on gene groups G_1 or G_2 independently. In each case, samples are clustered into two groups, which are marked as "a" or "b". We use green color (second line) to represent cluster results based on G_1 and blue color (third line) for results based on G_2. By combination, four possible sample groups are generated: C_1 includes samples marked as "a" based on G_1 and marked as "a" based on G_2; C_2 includes samples marked as "a" based on G_1 and marked as "b" based on G_2; C_3 includes samples marked as "b" based on G_1 and marked as "a" based on G_2; and C_4 includes samples marked as "b" based on G_1 and marked as "b" based on G_2.

If $k=3$, there will be eight possible sample groups. In general, the number of possible sample groups equals 2^k. Usually k is set to be 2 to reduce the computational complexity.

Step 4: finding heterogeneous groups. Among the sample groups C_1 to C_{2^k}, we choose two distinct groups C_s and C_t such that each sample in C_s is in the different cluster with each sample in C_t during clustering in Step 2. (C_s, C_t) is called a *heterogeneous group*. For example, if $k=2$, among the sample groups C_1, C_2, C_3, C_4, we choose two distinct groups C_s and C_t ($1 \leq s$, $t \leq 4$) which satisfy the following condition: for $\forall u \in C_s$, $\forall v \in C_t$, where u and v are samples, if $u \in S_{i,r1}$, $v \in S_{i,r2}$, then $r_1 \neq r_2$ (r_1, $r_2 \in \{a,b\}$) for all i ($1 \leq i \leq k$). (C_s, C_t) is then a heterogeneous group. For example, (C_1, C_4) is such a heterogeneous group (when $k = 2$) because all samples in group C_1 are clustered into $S_{i,a}$ ($1 \leq i \leq k$), while all samples in group C_4 are clustered into $S_{i,b}$ ($1 \leq i \leq k$). For the same reason, (C_2, C_3) is another heterogeneous group. We use these heterogeneous groups as the representation of the original sample partition.

Step 5: sorting and reducing. In this step, we reduce genes based on the sample patterns in the heterogeneous groups. To find genes whose expression patterns are strongly correlated with the class distinction within the heterogeneous group, we build on-off patterns according to the class distribution of each heterogeneous group and sort genes by their degree of correlation with the patterns. For example, for the heterogeneous group (C_1, C_4), two patterns $(0,0,...0,1,1,...1)$ and $(1,1,...1,0,0,...0)$ are introduced. The pattern $(0,0,...0,1,1,...1)$ includes $|C_1|$ (number of samples in group C_1) zeros followed by C_4 (number of samples in group C_4) one's. Similarly, $(1,1,...1,0,0,...0)$ includes $|C_1|$ one's followed by $|C_4|$ zeros. For each pattern, we use it to calculate vector-cosine defined in Equation (3) with each gene vector, then sort all genes according to the similarity values in descending order, and keep the first one third of the sorted gene sequence by cutting off the other two thirds of the gene sequence.[5] By merging the remaining sorted gene sequences from two patterns, we obtain the reduced gene sequence G' where at least one third of the genes in G are cut off.

Similarly, for the other heterogeneous group (C_2, C_3), another reduced gene sequence G'' is generated. Now the question is which gene subset should be chosen for the next iteration, G' or G''? The semantic meaning behind this is to select a heterogeneous group which is a better representation for the original distribution of samples because G' and G'' are generated based on the corresponding heterogeneous groups. Here we use the cross-validation method [Golub *et al.*, 1999] to evaluate each group. In each heterogeneous group, we first choose one sample and build on-off patterns using similar procedure mentioned above, but only based on the remaining samples. We then sort genes by their degree of correlation with the pattern to select important genes and predict the class of the withheld samples. The process is repeated for each sample, and the cumulative error rate is calculated. When the heterogeneous group which has lower error rate is found, its corresponding reduced gene sequence is selected as $\hat{G}$ with n_2 genes for the next iteration.

---

[5] This two thirds threshold is a trade-off of information preservation in each iteration and time cost of the whole approach.

Samples

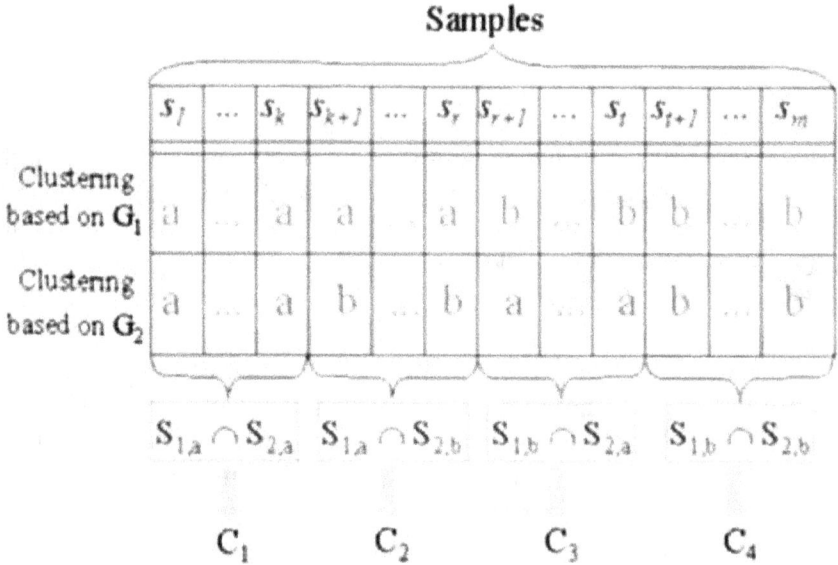

Figure 3. Clustering results combination when $k=2$. s_1, s_2, ... s_m in the first line represent samples.

After *Step 5*, the gene number is reduced from n_1 to n_2.

The above steps can be repeated by clustering n_2 genes, and so on. The iteration will be terminated until the termination conditions are satisfied, which is discussed in the following subsection.

8.2.4 Termination Condition

After one iteration involving detection of sample pattern and selection of genes, a certain number of genes will be pruned. The remaining genes and the entire samples then form a new gene expression matrix from which a new iteration starts.

We will now discuss the issue of determining when sufficient iterations have been performed. Ideally, iterations will be terminated when a stable and significant pattern of samples has emerged. Thus, the iteration termination criterion involves determining the measurement and threshold which identifies a "stable and significant" pattern.

The propose of clustering samples is based on identifying groups of empirical interesting patterns in the underlying samples. In general, we hope that the samples in a given group will be similar (or related) to one another and different from (or unrelated to) the samples in other groups. The greater the similarity (or homogeneity) within a group, and the greater the difference between groups, the better or more distinct the partition.

As described above, after each iteration, we use the remaining genes to classify samples and then use the *coefficient of variation (CV)* to measure how "internally-similar and well-separated" this partition is:

$$CV = \frac{1}{N}\sum_{k=1}^{N}\frac{\sigma_k}{\|\bar{\mu}_k\|}, \tag{5}$$

where N represents the cluster number, μ_k indicates the center of group k, and σ_k represents the standard deviation of group k. Assuming there are t objects $\{\vec{v}_1, \vec{v}_2, ..., \vec{v}_t\}$ in group k, each object is a n-dimensional vector $\vec{v}_j = <m_1, m_2, ...m_n>$. The center of group k is defined as:

$$\bar{\mu}_k = <\overline{m}_{1,k}, \overline{m}_{2,k}, ...\overline{m}_{n,k}>, \tag{6}$$

where

$$\overline{m}_{i,k} = \frac{1}{t}\sum_{j=1}^{t}m_{i,j}, \quad (i = 1,2,...,n) \tag{7}$$

And the standard deviation of group k is defined as:

$$\sigma_k = \frac{\sqrt{\sum_{i=1}^{t}\|\vec{v}_i - \vec{v}_k\|^2}}{t-1}. \tag{8}$$

It is clear that, if the data set contains an "internally-similar and well-separated" partition, the standard deviation of each group will be low, and the CV value is expected to be small. Thus, based on the coefficient of variation, we may conclude that small values of the index indicate the presence of a "good" pattern. In the interrelated two-way clustering approach, we examine the *coefficient of variation* values after each iteration and terminate the algorithm after an iteration with a CV value much smaller than the previous.

Another applicable termination condition involves checking whether the number of genes is small enough to guide sample class prediction. This number is highly dependent on the type of data. For example, in a typical biological system, the number of genes needed to fully characterize a macroscopic phenotype and the factors determining this number are often unclear. Experiments also show that, for certain data, gene numbers varying from *10~200* can all serve as good predictors [Golub *et al.*, 1999]. For our microarray data experiments, we have chosen *100* as a compromise termination number; e.g. when the number of genes falls below *100*, the iteration stops. This termination condition is used only as a supplementary criterion.

Genes that remain will be regarded as the selected genes resulting from this interrelated two-way clustering approach. They are then used to cluster the samples for a final result. Since the number of genes is relatively small, the traditional clustering methods can be applied to the selected genes. The remaining genes can also be treated as "predictors" to establish cluster labels such as disease symptoms and control condition for the next batch of samples.

8.3 Experimental Results

We will now present experimental results using four microarray data sets. The first two data sets are from a study of multiple-sclerosis patients collected by the Neurology and Pharmaceutical Sciences Departments of the State University of New York at Buffalo. Multiple sclerosis (MS) is a chronic, relapsing, inflammatory disease of the central nervous system that causes physical and cognitive disability in adults of working age between 16 and 60. Interferon-β (IFN-β) has offered the most important treatment for the MS disease over the last decade [Yong *et al.*, 1998]. The MS data set includes two groups: the MS_IFN group, containing 28 samples (14 MS, 14 IFN), and the CONTROL_MS group, containing 30 samples (15 MS, 15 Control). Each sample is measured over 4132 genes. The other two data sets are based on a collection of leukemia patient samples reported in (Golub *et al.*, 1999). One matrix includes 38 samples (27 ALL vs. 25 AML), and the other contains 34 samples (20 ALL, 14 AML). Each sample is measured over 7129 genes. The ground-truth of the partition, which includes such information as how many samples belong to each cluster and the cluster label for each sample, is used only to evaluate the experimental results.

During the data pre-processing procedure for MS_IFN group, by sorting gene vectors using vector-cosine calculated from Equation (3), we choose threshold *0.89* and then remove gene vectors for which vector-cosine with pattern E is higher than the threshold, which means that these gene intensity values vary little among the samples. Figure 4 shows the situation, where horizontal axis represents gene vectors and vertical axis represents vector-cosine values. Gene vectors are sorted in an ascending order, and then we choose threshold 0.89 to reduce 4132 genes to 2682. Thus, 1450 genes are removed from 4132. K-means clustering method is used during the interrelated two-way process, and correlation coefficient (Equation (4)) is used as the distance measure.

In Figure 5, a linear mapping function which maps the n-dimensional vectors onto two dimensions [Bhadra and Garg, 2001] is used to show the distribution of samples before and after the interrelated two-way clustering procedure. As indicated by this figure, prior to the application of the approach, the samples are uniformly scattered, with no obvious clusters. As the iterations proceed, sample clusters progressively emerge until in Figure 5(B), the samples are clearly separated into two groups. This visualization provides a clear illustration of the iterative process. The green and red dots indicate the clusters resulting from the interrelated two-way clustering

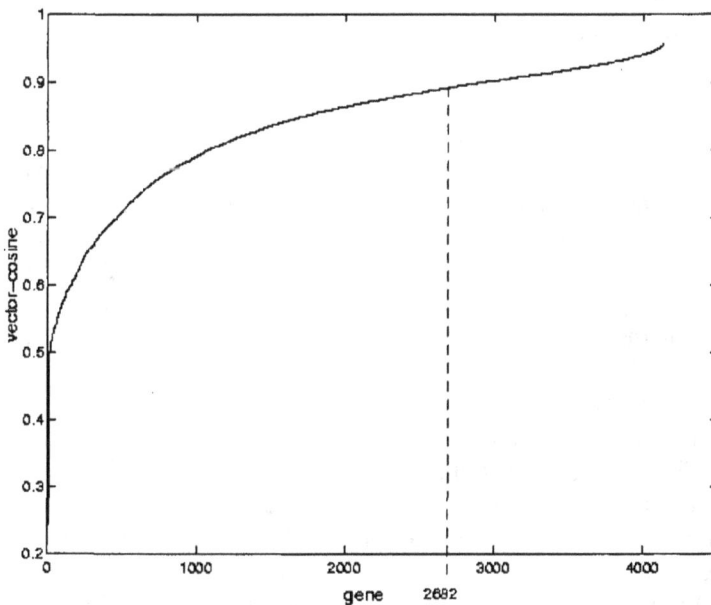

Figure 4. Distribution of gene vectors' vector-cosine calculated from Equation (3).

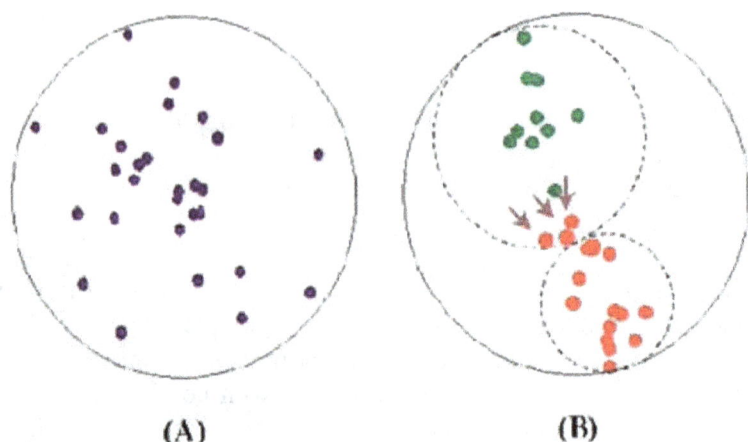

(A)   (B)

Figure 5. Clustering results on the MS_IFN group.

approach, while the two dashed circles show the actual partition of the samples, with arrows pointing out the incorrectly-classified samples. Here, the clustering approach selected 96 genes and classified 28 samples into two groups. 11 samples are in group one, matching the MS disease samples. Another 17 samples are in group two; of these, 14 are from the IFN treatment group and 3 are incorrectly matched.

Similarly, for the CONTROL_MS group, we removed 1474 genes using the same threshold as above for the MS_IFN group in the pre-processing step and performed the interrelated two-way cluster on the remaining 2658 genes. The result is eight samples being incorrectly classified out of 30 samples.

For the purpose of comparison, we also directly performed K-means clustering method and self-organizing maps on both the MS_IFN and CONTROL_MS groups data after normalization. These approaches are applied on both original gene data and the gene data after pre-processing. Figure 6 illustrates the sample clustering accuracy rate achieved by these methods. Figure 6(A) shows clustering results on the MS_IFN group which includes 28 samples. The first bar is the accuracy rate by ITC, the second bar is the accuracy rate by applying SOM on 4132 genes matrix, the third bar shows the result by applying SOM on 2682 genes matrix after pre processing, the forth bar shows the result by applying K-means on 4132 genes matrix,

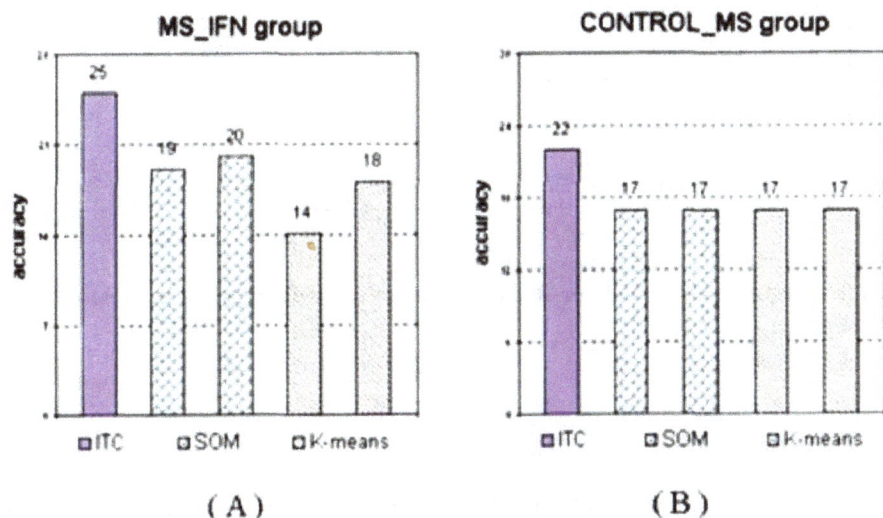

Figure 6. Comparison of accuracy rate achieved by interrelated two-way clustering (denoted as ITC), self-organizing maps (SOM), and K-means clustering methods.

and the last one shows the result by applying K-means on 2682 genes matrix. Figure 6(B) shows clustering results on the CONTROL_MS group which include 30 samples. The order of the rates shown is same as (A). Two bars for SOM and two bars for K-means as well are the accuracy rate on 4132 genes matrix and 2568 genes matrix after pre-processing, respectively.

From Figure 6, we can see that using our approach, the accuracy of class discovery is higher than those of traditional methods, which illustrates the effectiveness of the interrelated two-way clustering method on such high dimensional gene data.

We also applied interrelated clustering, self-organizing maps (SOM), and K-means clustering methods on two leukemia patient data sets. The accuracy rates reached by these algorithms are shown in Table 1. Note that we applied all three approaches on the original data sets, since the data they provided are already pre-processed. The performance of each method is measured by wrongly classified samples number (Error #). From the table, we can see that interrelated two-way clustering performs much better than the other two methods on these data sets.

Dataset	interrelated two-way	SOM	k-means
Error# of Set 1(38 samples)	5	10	15
Error# of Set 2(34 samples)	1	11	12

Table 1. Experiment results on leukemia patient samples reported in [Golub *et al.*, 1999].

8.4 Conclusion

In this chapter, we have presented a new framework for the unsupervised analysis of gene expression data. In this framework, an interrelated two-way clustering method is developed and applied on the gene expression matrices transformed from the raw microarray data. This approach can detect significant patterns within samples while dynamically selecting significant genes which manifest the conditions of actual empirical interest. We have shown that, during the iterative clustering, reducing genes can improve the accuracy of class discovery, which in turn will guide further genes reduction.

We have demonstrated the effectiveness of the above approach based on the experiments conducted on the multiple sclerosis data sets and two leukemia data sets. These experiments indicate that interrelated clustering appears to be a promising approach for unsupervised sample clustering on gene array data sets.

Acknowledgment

This research is supported by NSF grants.

References

Alizadeh, A.A., Eisen, M.B., Davis, R.E., Ma, C., Lossos, I.S., Rosenwald, A., Boldrick, J.C., Sabet, H., Tran, T., Yu, X., Powell, J.I., Yang, L., Marti, G.E., Moore, T., Hudson, J. Jr, Lu, L., Lewis, D.B., Tibshirani, R., Sherlock, G., Chan, W.C., Greiner, T.C., Weisenburger, D.D., Armitage, J.O., Warnke, R., Levy, R., Wilson, W., Grever, M.R., Byrd, J.C., Botstein, D., Brown, P.O. and Staudt, L.M. (2000) "Distinct types of diffuse large b-cell lymphoma identified by gene expression profiling." *Nature* **403,** 503-511.

Alon, U., Barkai, N., Notterman, D.A., Gish, K., Ybarra, S., Mack, D. and Levine, A.J. (1999) "Broad patterns of gene expression revealed by clustering analysis of tumor and normal colon tissues probed by oligonucleotide array." *Proc. Natl. Acad. Sci. USA* **96(12),** 6745-6750.

Alter, O., Brown, P.O. and Bostein, D. (2000) "Singular value decomposition for genome-wide expression data processing and modeling." *Proc. Natl. Acad. Sci. USA* **97(18),** 10101-10106.

Azuaje, F. (2000) "Making genome expression data meaningful: Prediction and discovery of classes of cancer through a connectionist learning approach." *Proceedings of IEEE International Symposium on Bioinformatics and Biomedical Engineering,* IEEE Computer Society Press, 208-213.

Barash, Y. and Friedman, N. (2001) "Context-specific bayesian clustering for gene expression data." *Proc. Fifth Annual International Conference on Computational Molecular Blology (RECOMB 2001),* ACM Press, 12-20.

Ben-Dor, A., Friedman, N. and Yakhini, Z. (2001) "Class discovery in gene expression data." *Proc. Fifth Annual International Conference on Computational Molecular Biology (RECOMB 2001),* ACM Press, 31-38.

Ben-Dor, A., Shamir, R. and Yakhini, Z. (1999) "Clustering gene expression patterns." *Journal of Computational Biology* **6(3/4),** 281-297.

Bhadra, D. and Garg, A. (2001) "An interactive visual framework for detecting clusters of a multidimensional dataset." Technical Report 2001-03, Dept. of Computer Science and Engineering, University at Buffalo, NY.

Brazma, A. and Vilo, J. (2000) "Minireview: Gene expression data analysis." *Federation of European Biochemical Societies* **480,** 17-24.

Brown, M.P.S., Grundy, W. N., Lin, D., Cristianini, N., Sugnet, C.W., Furey, T.S., Ares, M. Jr. and Haussler, D. (2000) "Knowledge-based analysis of microarray gene expression data using support vector machines." *Proc. Natl. Acad. Sci.* **97(1)**, 262-267.

Chen, J.J., Wu, R., Yang, P.C., Huang, J.Y., Sher, Y.P., Han, M.H., Kao, W.C., Lee, P.J., Chiu, T.F., Chang, F., Chu, Y.W., Wu, C.W. and Peck, K. (1998) "Profiling expression patterns and isolating differentially expressed genes by cDNA microarray system with colorimetry detection." *Genomics* **51**, 313-324.

DeRisi, J., Penland, L., Brown, P.O., Bittner, M.L., Meltzer, P.S., Ray, M., Chen, Y., Su, Y.A. and Trent, J.M. (1996) "Use of a cDNA microarray to analyse gene expression patterns in human cancer." *Nature Genetics* **14**, 457-460.

Devore, J.L. (1991) *Probability and Statistics for Engineering and Sciences.* Brook/Cole Publishing Company.

Eisen, M.B., Spellman, P.T., Brown, P.O. and Botstein, D. (1998) "Cluster analysis and display of genome-wide expression patterns." *Proc. Natl. Acad. Sci. USA* **95**, 14863-14868.

Ermolaeva, O., Rastogi, M., Pruitt, K.D., Schuler, G.D., Bittner, M.L., Chen, Y., Simon, R., Meltzer, P., Trent, J.M. and Boguski, M.S. (1998) "Data management and analysis for gene expression arrays." *Nature Genetics* **20**, 19-23.

Furey, T.S., Cristianini, N., Duffy, N., Bednarski, D.W., Schummer, M. and Haussler, D. (2000) "Support vector machine classification and validation of cancer tissue samples using microarray expression data." *Bioinformatics* **16(10)**, 909-914.

Getz, G., Levine, E. and Domany, E. (2000) "Coupled two-way clustering analysis of gene microarray data." *Proc. Natl. Acad. Sci. USA* **97(22)**, 12079-12084.

Golub, T.R., Slonim, D.K., Tamayo, P., Huard, C., Gassenbeek, M., Mesirov, J.P., Coller, H., Loh, M.L., Downing, J.R., Caligiuri, M.A., Bloomfield, D.D. and Lander, E.S. (1999) "Molecular classification of cancer: Class discovery and class prediction by gene expression monitoring." *Science* **286(15)**, 531-537.

Hakak, Y., Walker, J.R., Li, C., Wong, W.H., Davis, K.L., Buxbaum, J.D., Haroutunian, V. and Fienberg, A.A. (2001) "Genome-wide expression analysis reveals dysregulation of myelination-related genes in chronic schizophrenia." *Proc. Natl. Acad. Sci. USA* **98(8)**, 4746-4751.

Han, J. and Kamber, M. (2000) *Data Mining: Concept and Techniques.* The Morgan Kaufmann Series in Data Management Systems. Morgan Kaufmann Publishers.

Hartigan, J.A. (1975) *Clustering Algorithm.* John Wiley and Sons, New York.

Hartigan, J.A. and Wong, M.A. (1979) "Algorithm AS136: A k-means clustering algorithm." *Applied Statistics* **28**, 100-108.

Hastie, T., Tibshirani, R., Boststein, D. and Brown, P. (2001) "Supervised harvesting of expression trees." *Genome Biology* **2(1)**, 0003.1-0003.12.

Heller, R.A., Schena, M., Chai, A., Shalon, D., Bedilion, T., Gilmore, J., Woolley, D.E. and Davis, R.W. (1997) "Discovery and analysis of inflammatory disease-related genes using cDNA microarrays." *Proc. Natl. Acad. Sci. USA* **94**, 2150-2155.

Herrero, J., Valencia, A. and Dopazo, J. (2001) "A hierarchical unsupervised growing neural network for clustering gene expression patterns." *Bioinformatics* **17**, 126-136.

Holter, N.S., Mitra, M., Maritan, A., Cieplak, M., Banavar, J.R. and Fedoroff, N.V. (2000) "Fundamental patterns underlying gene expression profiles: simplicity from complexity." *Proc. Natl. Acad. Sci. USA* **97(15)**, 8409-8414.

Iyer, V.R., Eisen, M.B., Ross, D.T., Schuler, G., Moore, T., Lee, J.C.F., Trent, J.M., Staudt, L.M., Hudson, J. Jr., Boguski, M.S., Lashkari, D., Shalon, D., Botstein, D. and Brown, P.O. (1999) "The transcriptional program in the response of human fibroblasts to serum." *Science* **283**, 83-87.

Jiang, S., Tang, C., Zhang, L., Zhang, A. and Ramanathan, M. (2001) "A maximum entropy approach to classifying gene array data sets." *Proc. of Workshop on Data Mining for Genomics, First SIAM International Conference on Data Mining.*

Jorgensen, A. (2000) "Clustering excipient near infrared spectra using different chemometric methods." Technical Report, Dept. of Pharmacy, University of Helsinki.

Kohonen, T. (1984) *Self-Organization and Associative Memory.* Spring-Verlag, Berlin.

Manduchi, E., Grant, G.R., McKenzie, S.E., Overton, G.C., Surrey, S. and Stoeckert, C.J. Jr. (2000) "Generation of patterns from gene expression data by assigning confidence to differentially expressed genes." *Bioinformatics* **16(8)**, 685-698.

Martin, K.J., Graner, E., Li, Y., Price, L.M., Kritzman, B.M., Fournier, M.V., Rhei, E. and Pardee, A.B. (2001) "High-sensitivity array analysis of gene expression for the early detection of disseminated breast tumor cells in peripheral blood." *Proc. Natl. Acad. Sci. USA* **98(5)**, 2646-2651.

Mody, M., Cao, Y., Cui, Z., Tay, K.Y., Shyong, A., Shimizu, E., Pham, K., Schultz, P., Welsh, D. and Tsien, J.Z. (2001) "Genome-wide gene expression profiles of the developing mouse hippocampus." *Proc. Natl. Acad. Sci. USA* **98(15)**, 8862-8867.

Moler, E.J., Chow, M.L. and Mian, I.S. (2000) "Analysis of molecular profile data using generative and discriminative methods." *Physiological Genomics* **4(2)**, 109-126.

Park, P.J., Pagano, M. and Bonetti, M. (2001) "A nonparametric scoring algorithm for identifying informative genes from microarray data." *Pacific Symposium on Biocomputing*, 52-63.

Pavlidis, P., Weston, J., Cai, J. and Grundy, W.N. (2001) "Gene functional classification from heterogeneous data." *RECOMB 2001: Proceedings of the Fifth Annual International Conference on Computational Biology*, ACM Press, 249-255.

Perou, C.M., Jeffrey, S.S., Rijn, M.V.D., Rees, C.A., Eisen, M.B., Ross, D.T., Pergamenschikov, A., Williams, C.F., Zhu, S.X., Lee, J.C.F., Lashkari, D., Shalon, D., Brown, P.O. and Bostein, D. (1999) "Distinctive gene expression patterns in human mammary epithelial cells and breast cancers." *Proc. Natl. Acad. Sci. USA* **96(16),** 9212-9217.

Schena, M., Shalon, D., Davis, R.W. and Brown, P.O. (1995) "Quantitative monitoring of gene expression patterns with a complementary DNA microarray." *Science* **270,** 467-470.

Schena, M., Shalon, D., Heller, R., Chai, A., Brown, P.O. and Davis, R.W. (1996) "Parallel human genome analysis: Microarray-based expression monitoring of 1000 genes." *Proc. Natl. Acad. Sci. USA* **93(20),** 10614-10619.

Schuchhardt, J., Beule, D., Malik, A., Wolski, E., Eickhoff, H., Lehrach, H. and Herzel, H. (2000) "Normalization strategies for cDNA microarrays." *Nucleic Acids Research* **28(10)**.

Shalon, D., Smith, S.J. and Brown, P.O. (1996) "A DNA microarray system for analyzing complex DNA samples using two-color fluorescent probe hybridization." *Genome Research* **6**, 639-645.

Slonim, D.K., Tamayo, P., Mesirov, J.P., Golub, T.R. and Lander, E.S. (2000) "Class prediction and discovery using gene expression data." *RECOMB 2000: Proceedings of the Fifth Annual International Conference on Computational Biology*. ACM Press.

Spellman, P.T., Sherlock, G., Zhang, M.Q., Iyer, V.R., Anders, K., Eisen, M.B., Brown, P.O., Botstein, D. and Futcher, B. (1998) "Exploring the metabolic and genetic control of gene expression on a genomic scale." *Mol. Biol. Cell,* 3273.

Tamayo, P., Solni, D., Mesirov, J., Zhu, Q., Kitareewan, S., Dmitrovsky, E., Lander, E. S. and Golub, T.R. (1999) "Interpreting patterns of gene expression with self-organizing maps: Methods and application to hematopoietic differentiation." *Proc. Natl. Acad. Sci. USA* **96(6),** 2907-2912.

Tavazoie, S., Hughes, D., Campbell, M.J., Cho, R.J. and Church, G.M. (1999) "Systematic determination of genetic network architecture." *Nature Genet.,* 281-285.

Thomas, J.G., Olson, J.M., Tapscott, S.J. and Zhao, L.P. (2001) "An efficient and robust statistical modeling approach to discover differentially expressed genes using genomic expression profiles." *Genome Research* **11(7),** 1227-1236.

Virtaneva, K., Wright, F.A., Tanner, S.M., Yuan, B., Lemon, W.J., Caligiuri, M.A., Bloomfield, C.D., de La Chapelle, A. and Krahe, R. (2001) "Expression profiling reveals fundamental biological differences in acute myeloid leukemia with isolated trisomy 8 and normal cytogenetic." *Proc. Natl. Acad. Sci. USA* **98(3),** 1124-1129.

Welford, S.M., Gregg, J., Chen, E., Garrison, D., Sorensen, P.H., Denny, C.T. and Nelson, S.F. (1998) "Detection of differentially expressed genes in primary tumor tissues using representational differences analysis coupled to microarray hybridization." *Nucleic Acids Research* **26,** 3059-3065.

Welsh, J.B., Zarrinkar, P.P., Sapinoso, L.M., Kern, S.G., Behling, C.A., Monk, B.J., Lockhart, D.J., Burger, R.A. and Hampton, G.M. (2001) "Analysis of gene expression profiles in normal and neoplastic ovarian tissue samples identifies candidate molecular markers of epithelial ovarian cancer." *Proc. Natl. Acad. Sci. USA* **98(3),** 1176-1181.

Xing, E.P. and Karp, R.M. (2001) "Cliff: Clustering of high-dimensional microarray data via iterative feature filtering using normalized cuts." *Bioinformatics* **17(1),** 306-315.

Yang, Y.H., Dudoit, S., Luu, P. and Speed, T.P. (2001) "Normalization for cDNA microarray data." *Proceedings of SPIE BiOS 2001: The International Society for Optical Engineering, International Biomedical Optics Symposium,* San Jose, California.

Yong, V., Chabot, S., Stuve, Q. and Williams, G. (1998) "Interferon beta in the treatment of multiple sclerosis: Mechanisms of action." *Neurology* **51**, 682-689.

Zhang, H., Yu, C.Y., Singer, B. and Xiong, M. (2001) "Recursive partitioning for tumor classification with gene expression microarray data." *Proc. Natl. Acad. Sci. USA* **98(12)**, 6730-6735.

Zou, S., Meadows, S., Sharp, L., Jan, L.Y. and Jan, Y. N. (2000) "Genome-wide study of aging and oxidative stress response in *Drosophila Melanogaster*." *Proc. Natl. Acad. Sci. USA* **97(25)**, 13726-13731.

Authors' Addresses

Chun Tang, Department of Computer Science and Engineering, State University of New York at Buffalo, Buffalo, NY 14260, USA. Email: chuntang@cse.buffalo.edu.

Li Zhang, Department of Computer Science and Engineering, State University of New York at Buffalo, Buffalo, NY 14260, USA. Email: lizhang@cse.buffalo.edu.

Aidong Zhang, Department of Computer Science and Engineering, State University of New York at Buffalo, Buffalo, NY 14260, USA. Email: azhang@cse.buffalo.edu.

Murali Ramanathan, Department of Pharmaceutical Sciences, State University of New York at Buffalo, Buffalo, NY 14260, USA. Email: murali@acsu.buffalo.edu.

Chapter 9

Creating Metabolic Network Models using Text Mining and Expert Knowledge

J.A. Dickerson, D. Berleant, Z. Cox, W. Qi, D. Ashlock, E.S. Wurtele and A.W. Fulmer

9.1 Introduction

RNA profiling analysis and new techniques such as proteomics (the profiling of proteins) and metabolomics (the profiling of small molecules) are yielding vast amounts of data on gene expression. This points to the need to develop new methodologies to identify and analyze complex biological networks. This chapter describes the development of a Java™-based tool that helps dynamically find and visualize metabolic networks. The tool consists of three parts. The first part is a text-mining tool that pulls out potential metabolic relationships from the PubMed database. These relationships are then reviewed by a domain expert and added to an existing network model. The result is visualized using an interactive graph display module. The basic metabolic or regulatory flow in the network is modeled using fuzzy cognitive maps. Causal connections are pulled out from sequence data using a genetic algorithm-based logical proposition generator that searches for temporal patterns in microarray data. Examples from the regulatory and metabolic network for the plant hormone gibberellin show how this tool operates.

The goal of this project is to develop a publicly available software suite called the Gene Expression Toolkit (GET). This toolkit will aid in the analysis and comparison of large microarray, proteomics and metabolomics data sets. It also aids in the synthesis of the new test results into the existing body of knowledge on metabolism. The user can select parameters for comparison such as species, experimental conditions, and developmental stage. The key tools in the Gene Expression Toolkit are:

- **PathBinder: Automatic document processing system** that mines online literature and extracts candidate relationships from publication abstracts.
- **ChipView: Explanatory models** synthesized by clustering techniques together with a genetic algorithm-based data-mining tool.
- **FCModeler: Predictive models** summarize known metabolic relationships in fuzzy cognitive maps (FCMs).

Figure 1 shows the relationship between the different modules. The PathBinder citations are available to the researcher and smoothly transferable for use in annotating displays in other parts of the package and as links in building models. ChipView searches for link hypotheses in microarray data. The FCModeler tool for gene regulatory and metabolic networks is intended to easily capture the intuitions of biologists and help test hypotheses along with providing a modeling framework for putting the results of large microarray studies in context.

9.2 Structure of Concepts and Links

The nodes in the metabolic network represent specific biochemicals such as proteins, RNA, and small molecules (metabolites), or stimuli, such as light, heat, or nutrients. Three basic types of directed links are specified: conversion, regulatory, and catalytic. In a conversion link (black arrow, shown as a heavy dotted line), a node (usually representing a chemical) is converted into another node, and used up in the process. In a regulatory link (green and red arrows, shown as solid and dashed arrows respectively), the node activates or deactivates another node, and is not used up in the process.

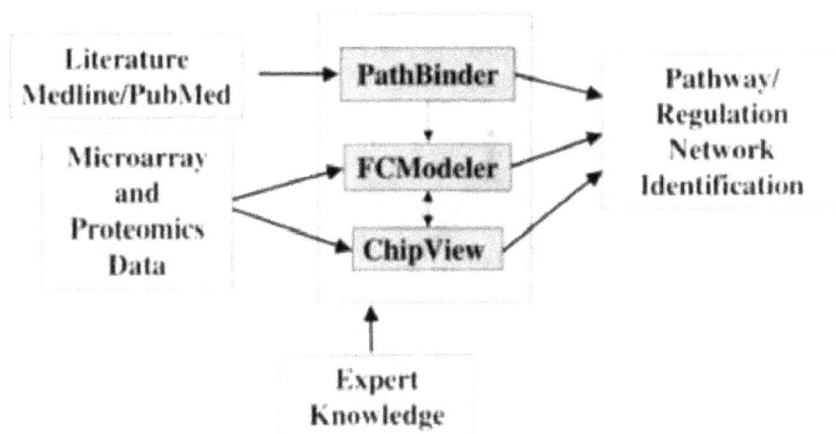

Figure 1. The Gene Expression Toolkit consists of PathBinder, FCModeler, and ChipView. The inputs to the system are the literature databases such as PubMed, experimental results from RNA microarray experiments, proteomics, and the expert knowledge and experience of the biologists that study an organism. The result will be a predictive model of the metabolic pathways.

A catalytic link (blue arrows, shown as a thick line) represents an enzyme that enables a chemical conversion and does not get used up in the process. Figure 2 shows a small part of a graph for the Arabidopsis metabolic and regulatory network. There is also an undirected link that defines a connection between two nodes and does not specify a direction of causality.

In the metabolic network database, the type of link is further delineated by the link mechanism and the certainty. Some of the current mechanisms are: direct, indirect, and ligand. Direct links assume a direct physical interaction. Indirect links assume that the upstream node activates the downstream node indirectly and allows for the existence of intermediate nodes in such a path. The ligand link is a "second messenger" mechanism in which a node produces or helps produce a ligand (small molecule that binds) and either "activates" or "inhibits" a target node. Often the nature of the link is unknown and it cannot be modeled in the current framework. The link certainty expresses a degree of confidence about the link. This will be used for hypothesis testing.

Other key features include concentrations of the molecules (nodes), strengths of the links, and subcellular compartmentation. These data can be

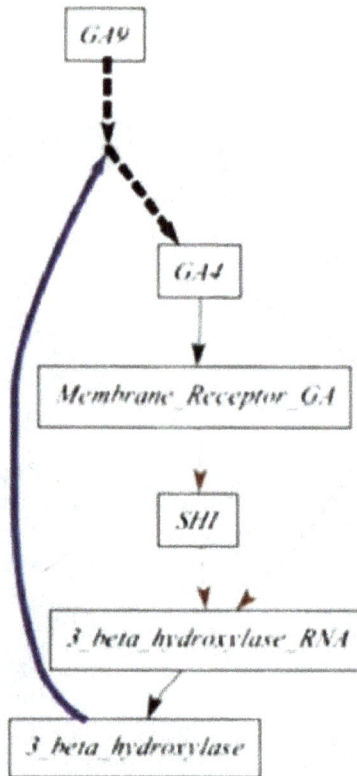

Figure 2. This is a map of a simple metabolic model of gibberellin (active form is GA4). The sequence is started by translation of 3_beta_ hydroxylase_RNA into the 3_beta_ hydroxylase protein. Bold dashed lines are conversion links, bold lines are catalytic links, thin solid lines are positive regulatory links and dashed thin lines are negative regulatory links.

added as they are identified experimentally. Currently the biologist user can include or ignore a variety of parameters, such as subcellular compartmentation and link strength. Since the node and link data is entered into a relational database, individual biologists can easily sort, share, and post data on the web. Future versions will distinguish between regulation that results in changes in concentrations of the regulated molecule, and regulation that involves a reversible activation or deactivation.

9.3 PathBinder: Document Processing Tool

PathBinder identifies information about the pathways that mediate biological processes from the scientific literature. This tool searches through documents in MEDLINE for passages containing terms that indicate relevance to signal transduction or metabolic pathways of interest. Microarray data can be used to hypothesize causal relationships between genes. PathBinder then mines MEDLINE for information about these putative pathways, extracting passages most likely to be relevant to a particular pathway and storing this desired information. The information is presented in a user-friendly format that supports efficiently investigating the pathways.

Related Work on Knowledge Extraction from Biochemistry Literature

An increasing body of work addresses extraction of knowledge from biochemical literature. Some works compare documents, such as MEDLINE abstracts, and extract information from the comparisons. For example, Shatkay *et al.* and Stapley assess the relatedness of genes based on the relatedness of texts in which they are mentioned [Shatkay, 2000; Stapley, 2000]. Shatkay *et al.* get documents containing a particular gene, compare the set of documents to the set relevant to other genes, and if two sets are similar then the two genes are deemed related. Stapley compares the literatures of two genes and assesses relatedness of genes based on the rate at which papers contain both of them. The system presented by Usuzaka *et al.* learns to retrieve relevant abstracts from MEDLINE based on examples of known relevant articles [Usuzaka, 1998].

Other works directly address the relationships among entities such as proteins, genes, drugs, and diseases. An initial requirement for such a system is identifying relevant nouns. This can be done by extracting names from free text based on their morphological properties. Sekimizu *et al.* [1998] parse text to identify noun phrases, rather than concentrating on the nouns themselves. The GENIA system and the PROPER system address the need to identify relevant terms automatically to enable automatic maintenance of lexicons of proteins and genes [Fukuda *et al.*, 1998; Collier, 1999]. Proux *et al.* [1998] concentrate on gene names and symbols.

Once the lexicon problem has been addressed, text can be analyzed to extract relationships among entities discussed therein. Andrade and Valencia

[1998] extract sentences that contain information about protein function. Rindflesch *et al.* [1999] concentrate specifically on binding relationships (among macromolecules). Rindflesch *et al.* [2000] emphasize drug-gene-cell relationships bearing on cancer therapy. Thomas *et al.* [2000] use automatic protein name identification to support automatic extraction of interactions among proteins. Sekimizu *et al.* [1998] use automatically identified relevant noun phrases in conjunction with a hand-generated list of verbs to automatically identify subject-verb-object relationships stated in texts in MEDLINE. Craven and Kumlien [1999] extract relationships between proteins and drugs. They investigate two machine learning techniques in which a hand-classified training set is given to the system, which uses this set to infer criteria for deciding if other passages describe the relevant relationships. One machine learning technique is based on modeling passages as unordered sets of words, and assumes word co-occurrence probabilities are independent of one another (the Naïve Bayes approach). Tanabe *et al.* [1999] extract relationships between genes and between genes and drugs. Their MedMiner system supports human literature searches by retrieving and serving sentences from abstracts on MEDLINE over the Web, based on their keyword content. MedMiner is tuned to finding relationship-relevant sentences in abstracts that contain a gene name and relationship keyword, pair of gene names and relationship keyword, or a gene and a drug name and relationship keyword. MedMiner can also handle Boolean queries, such as those containing two protein names. In such cases MedMiner takes a query consisting of an OR'ed list of "primary" terms and an AND'ed list of "secondary" terms. A returned sentence must contain a "primary" term and a relationship word. Relationship words are from a relatively large lexicon of such terms predefined by the system.

A number of works address extracting relationships among proteins from biochemical texts. A solution enables both automatic construction of biochemical pathways, and assistance to investigators in identifying relevant information about proteins of interest to them.

Humphreys *et al.* [2000] specifically address enzyme reactions extracted from *Biochimica et Biophysica Acta* and *FEMS Microbiology Letters*. Such interactions are intended to support metabolic network construction. Rindflesch *et al.* [1999] apply non-trivial natural language processing (NLP) to extract assertions about binding relationships among proteins. Noun phrases are identified by a sophisticated combination of text processing and reference to existing name repositories.

Other systems have been reported that extract many interactions among diverse proteins. Blaschke *et al.* [1999] extract such interactions by first identifying phrases conforming to the template `protein... verbclass ...protein`, where `verbclass` is one of 14 sets of pathway relevant verbs (such as "bind") and their inflections. Protein names and synonyms are provided as an input and sentences containing extracted phrases are returned. The BioNLP subsystem, a component of a larger system, extracts sentences containing pathway relevant verbs determined by the user and applies templates to them to identify path relevant relationships among proteins [Ng, 1999; Wong, 2001]. Protein names are determined automatically. The subsystem, CPL2Perl, thresholds the results so that it ignores interactions with a single relevant sentence. This is useful if the sentence analysis was mistaken. Such a thresholding strategy tends to increase precision at the expense of reducing recall. Thomas *et al.* [2000] distinguish between verbs that are relatively more and less reliable in indicating protein interactions. Their system automatically recognizes protein names and relies on the strategy of tuning an existing sophisticated general-purpose natural language processing system to the protein interaction domain. Ono *et al.* [2001] use part-of-speech (POS) tagging, key verbs, and template matching on phrases to extract protein-protein interactions. Their system has an information retrieval effectiveness measure of up to 0.89 [Ding *et al.*, 2002].

PathBinder Operation

The PathBinder system, like previous works, extracts relevant passages about protein relationships from MEDLINE. The PathBinder work differs from these due to a combination of system design decisions. PathBinder avoids syntactic analysis of text in favor of word experts for pathway relevant verbs. Word experts are sets of rules for interpreting words [Berleant, 1995]. PathBinder also is oriented toward assisting humans in constructing pathways rather than fully automatic construction, thus avoiding some information retrieval precision limitations. We are also investigating the relative performances of several algorithms for identifying relevant sentences, including verb-free algorithms that rely instead on protein term co-occurrences. PathBinder relies on the sentence unit rather than abstracts, phrases, or other units because sentences rate highly on information retrieval effectiveness under reasonable conditions [Ding *et al.*, 2002].

How PathBinder Works

Step 1: user input. Keyboard input of biomolecule names in pathways of interest by the user.

Step 2: synonym extraction. A user-editable synonym file is combined with a more advanced module that will automatically access the HUGO (http://www.gene.ucl.ac.uk.publicfiles/nomen/nomenclature.txt) and OMIM (www.ncbi.nlm.nih.gov/htbinpost/Omim/) nomenclature databases, and extract synonyms.

Step 3: document retrieval. PubMed is accessed and queried using terms input in Step 1. The output of this step is a list of URLs with high relevance probabilities.

Step 4: sentence extraction. Each URL is downloaded and scanned for pathway-relevant sentences that satisfy the query. These sentences constitute pathway-relevant information "nuggets."

Repetition of steps 2 through 4, using different biomolecule names extracted from qualifying sentences. These new biomolecule names are candidates for inclusion in the pathways of interest.

Step 5: sentence index. Process the collection of qualifying sentences into a more user-friendly form, a multi-level index (Figure 3), with the number of levels dependent on the sentence extraction criteria. This index conforms to a pattern, displayed by a Web browser, and the sentences in it are clickable. When a sentence is clicked, the document from which it came appears in the Web browser.

Step 6: integration with the rest of the software and the microarray data sets. The index can be used to create a graphical representation in which verbs are represented by lines, interconnecting the biomolecule names and forming a web-like relationship diagram of the extracted information.

PathBinder is useful as both a standalone tool and an integrated subsystem of the complete system. The multilevel indexes transform naturally into inputs for the network modeling tools. The networks that PathBinder helps identify will form valuable input to the clustering, display, and analysis software modules.

```
┌─────────────────────────────────────────────┐
│  Protein A                                    │
│      Protein B                                │
│          Associates/Associated/etc.           │
│              Sentence 1                        │
│              Sentence 2                        │
│                                                │
│              . . .                             │
│                                                │
│          Binds/Binding/Bind/etc.               │
│              Sentence M                        │
│              Sentence M+1                      │
│                                                │
│              . . .                             │
│                                                │
│          Regulates/Regulating/etc.             │
│              . . .                             │
│      Protein C                                 │
│          Associates/Associated/etc.            │
│              Sentence M+N                       │
│                                                │
│              . . .                             │
│                                                │
│          Binds/Binding/Bind/etc.               │
│              . . .                             │
│  Protein B                                     │
│      Protein D                                 │
│          Associates/Associated/etc.            │
│              Sentence M+N+P                     │
└─────────────────────────────────────────────┘
```

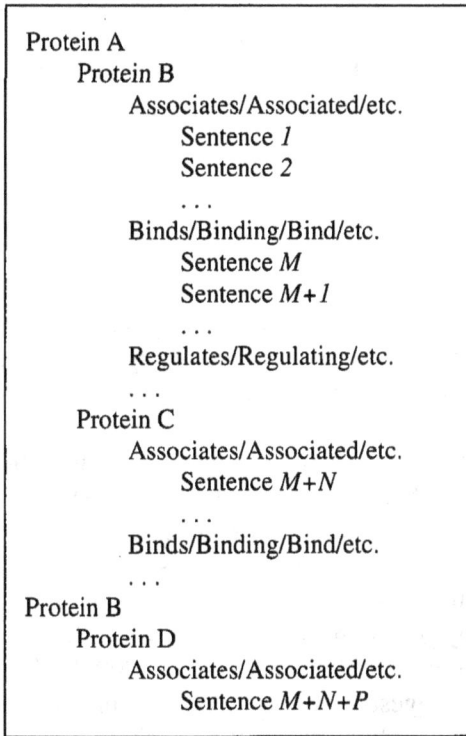

Figure 3. The long and somewhat disorganized sentence set that PathBinder extracts is converted into a multilevel index which is more suited to a human user. "Protein A", "Protein B", etc. are placeholders for the actual name of a path-relevant protein, and "Sentence 1", "Sentence 2", etc. are placeholders that would be actual sentences in the PathBinder-generated index.

Example of a Sample PathBinder Query

The query is to find sentences containing (either gibberellin, gibberellins, or GA) AND (either SPY, SPY-4, SPY-5, or SPY-7). Three relevant results were found and incorporated into the metabolic and regulatory visualization. A single sentence example is shown below.

Sentence: "The results of these experiments show that spy-7 and gar2-1 affect the GA dose-response relationship for a wide range of GA responses and suggest that all GA-regulated processes are controlled through a negatively acting GA-signaling pathway."

Source Information: UI—99214450, Peng J, Richards DE, Moritz T, Cano-Delgado A, Harberd NP, Plant Physiol 1999 Apr; 119(4):1199-1208.

9.4 ChipView: Logical Proposition Generator

Gene expression data is gathered as a series of snapshots of the expression levels of a large number of genes. The snapshots may be organized as a time series or a sequence of organism states. When multiple gene expression experiments are performed, the choice of genes, time points, or organism states often varies. Finally, the data gathered often contain many unusable points for a number of reasons. The variation in which data is collected, the noisy character of the data, and the fact that data is often missing mean that a gene expression analysis tool must be designed with all these limitations in mind. Current analysis tools, mostly built around clustering of various sorts, are quite valuable in cutting through the thickets of data generated by gene expression technology to find nuggets of truth (see for example [Eisen *et al.*, 1998; Brown *et al.*, 2000]). These tools, however, do not currently suggest possible interpretations to the researcher and incorporate many ad hoc assumptions about the mathematical and algorithmic behavior of various clustering techniques.

One possible way of addressing both the data collection limitations and lack of theoretical foundation is the Logical Proposition Generator. The key features of this tool are:

- Filtration of data items by behavioral abstractions that yield both interpretation of data and partial resistance to variations in data collection.
- Incorporation of a vast space of clustering techniques into the tool to create data driven, problem-specific clustering on the fly.
- Designing the tool so that its basic data objects are logical propositions about the data it is working with.

This makes the analogy to clustering in the logical proposition generator one that transparently supplies multiple potential interpretations of the data. The output of the tool is in the form of logical sentences with atoms drawn from absolute and differential classifications of expression profiles and relative abstractions of pairs of gene expression profiles. The prototype tool was written for gene expression profiles that are time series. Extending the logical proposition generator to have logical primitives that are appropriate for non-time series data is one of the goals as well.

Operation of the Logical Proposition Generator

Let us now specify the atoms and connective of the logical proposition language that is the target of the tool's search of the data for meaning. The tool permits the user to specify the expression level E that they believe specifies up or down regulation of a gene and the minimum change in expression level D that represents a significant change between adjacent time points. The tool recognizes classes of expression profiles given by the regulation state at each time point. Thus, "up, not down, not unchanged, down, down, not up, unchanged," specifies one of the possible classes of a seven point time series. Likewise, if +/− means significant change up or down since the last time step "+++00 − −" would represent a class of profiles that first increased, then stayed level, and later decreased their regulation between time steps. These two types of classes of expression profiles form the single expression profile atoms of the language.

The tool also uses logical atoms that compare pairs of profiles. These compute representative facts about the profiles, such as "profile one has its maximum before profile two", "the maximum change in regulation of the second profile exceeds that of the first", or "upregulation in the first profile does not occur unless a change in regulation has occurred in the second". The absolute and differential (single expression profile) atoms and the relative (two expression profile) atoms both return a "true" or "false" result. With these atoms available we then use traditional Boolean connectives AND, OR, NOT, XOR, etc. to build logical propositions.

Once we have the ability to make logical statements about gene expression profiles, the problem then becomes locating interesting and informative propositions. Statements that are always true, tautologies, are not interesting. Instead, we use a form of evolutionary computation, genetic programming [Koza, 1992; Kinnear, 1994; Koza, 1994; Angeline, 1996] to locate propositions that are true of subsets of the expression profiles. While this can be done blindly, with utility similar to clustering, it is also possible to force the expressions to be true when one of their arguments comes from a restricted class of genes of interest, e.g. a class we are trying to modify the expression of by some intervention. Thus, to find genes important to the upregulation of a class of genes X, we would search for propositions $P[x, y]$ that are often true when x is in X, seldom true when x was not in X, for some substantial but **not** universal collection Y of values for y. These vague statements about "usually true" and "substantial" become mathematically precise when embedded into the evolutionary search tool as a fitness function. One target of the research is an understanding of which fitness functions among those possible provide results useful to biological researchers.

The   relation $\{x \in 2233333\} \wedge \{y \in 5566666\} \wedge \{x \text{ first up before } y\}$ defines a binary relation of expression profiles. x must **not** change significantly at first while y must change at first. Later, x **must not go down** while y **must not go up** OR the first significant upregulation of x **must be before** that of y. Evolving such expressions permits the computation of interesting hypotheses about relations between profiles *including* relationships that use edges in the graphical models.

The logical proposition generator, by working with abstractions of the data in the form of the logical atoms described above yields the advantage that it is resistant, though certainly not immune, to variations in exactly which data are collected. The absolute and differential expression classes represent primitive fragments, which Boolean operations fuse together into data partitions, i.e. clusters. This means that the clustering techniques required to make sense of gene expression data are incorporated transparently into the logical proposition generator. Finally, in addition to locating genes that are implicated in the regulation of genes of interest, something clustering tools can do to some degree, the logical character of the tool will sometimes simultaneously suggest the "what" or "why" of the relationship, easing the work of interpretation and providing a source of tentative links for the other tools. This tool is not intended to replace clustering tools but to complement them. One way to locate a target set of genes, for example, might be to choose a tight cluster containing a few genes of interest and use this as a group of interest for the logical proposition generator.

Code	Measurement Change
1	Upregulated
2	Didn't change significantly
3	Didn't downregulate
4	Downregulated
5	Changed significantly from the baseline
6	Didn't upregulate
7	Matches anything

Table 1. Codes for changes in the expression profiles.

Example of Logical Proposition Generator Operation

The logical proposition generator operates on sets of expression profiles. It characterizes desired sequences as a series of numbers, e.g. Y in L: 124 means that Y is in the set of profiles that are in the state "Upregulated, didn't change, and downregulated". Table 1 gives the codes used in this example. An example logical proposition is given below:

```
(NAND
        (NOR
                (Y in L: 757243126155)
                (NAND (SamePro Y X) F))
                (AND T (NOT (NOT (NOR F T)))
        )
)
```

This is a logical proposition that acts on two 12-time-point expression profiles X and Y. It uses the logical operations *NAND*, *NOR*, *NOT*, and *AND* and the constants T and F. The logical proposition uses the binary predicate

"*SamePro*" which is true if two profiles are significantly up-and-down regulated in the same pattern. It also uses the unary predicate "*Y* in *L*:757243126155" which tests to see if *Y* is in the class of profiles that displays a particular pattern of up and down regulation in its twelve time points according to the scheme in Table 1.

Logical propositions of this form have the potential to encode very complex classes of expression profiles in very short statements. The following logical proposition also uses *OR* and *Say*, which we use to encode the logical identity, as well as differential classes, e.g. "*X* in *D*:73512467452" which check for changes in regulation since the last time step rather than as compared to the baseline:

(*NOR* (*Say* (*X* in *D*:73512467452))
   (*Say* (*OR* (*OR* (*X* in *D*:71661716551) (*X* in *L*:177621456644))
     (*NAND T* (*Say* (*Y* in *D*:13376357161))))
)
)

The *Say* operation does nothing but it leaves space in an expression that makes it easier for the evolutionary training techniques we use to move around sub-expressions that form coherent logical units.

9.5 Fuzzy Cognitive Map Modeling Tool for Metabolic Networks

The FCModeler tool for gene regulatory and metabolic networks captures the known metabolic information and expert knowledge of biologists in a graphical form. The node and link data for the metabolic map is stored in a relational database. This tool uses fuzzy methods for modeling network nodes and links and interprets the results using fuzzy cognitive maps [Kosko, 1986a; Kosko, 1986b; Dickerson and Kosko, 1994]. This tool concentrates on dynamic graphical visualizations that can be changed and updated by the user. This allows for hypothesis testing and experimentation.

Metabolic Network Mapping Projects

Two existing projects for metabolic networks are the Kyoto Encyclopedia of Genes and Genomes [Kanehisa and Goto, 2000] (KEGG http://www.genome.ad.jp/kegg) and the WIT Project [Overbeek *et al.*, 2000] (http://wit.mcs.anl.gov/WIT2/WIT). The WIT Project produces "metabolic reconstructions" for sequenced (or partially sequenced) genomes. It currently provides a set of over 39 such reconstructions in varying states of completion from the Metabolic Pathway Database constructed by Evgeni Selkov and his team. A metabolic reconstruction is a model of the metabolism of the organism derived from sequence, biochemical, and phenotypic data. This work is a static presentation of the metabolism asserted for an organism. The purpose of KEGG is to computerize current knowledge of molecular and cellular biology in terms of the information pathways that consist of interacting genes or molecules and, second, to link individual components of the pathways with the gene catalogs being produced by the genome projects. These metabolic reconstructions form the necessary foundation for eventual simulations.

E-CELL is a model-building kit: a set of software tools that allows a user to specify a cell's genes, proteins, and other molecules, describe their individual interactions, and then compute how they work together as a system [Tomita *et al.*, 1997; Tomita *et al.*, 1999; Tomita, 2001]. Its goal is to allow investigators to conduct experiments "in silico." Tomita's group has used versions of E-CELL to construct a hypothetical cell with 127 genes based on data from the WIT database. The E-CELL system allows a user to define a set of reaction rules for cellular metabolism. E-CELL simulates cell behavior by numerically integrating the differential equations described implicitly in these reaction rules.

EcoCyc is a pathway/genome database for *Escherichia coli* that describes its enzymes, and its transport proteins [Karp *et al.*, 2000] (http://ecocyc.DoubleTwist.com/ecocyc/). MetaCyc is a metabolic-pathway database that describes pathways and enzymes for many different organisms. These functional databases are publicly available on the web. The databases combine information from a number of sources and provide function-based retrieval of DNA or protein sequences. Combining this information has aided in the search for effective new drugs [Karp *et al.*, 1999]. EcoCyc has also made significant advances in visualizing metabolic pathways using stored layouts and linking data from microarray tests to the pathway layout [Karp *et al.*, 1999].

Visualizing Metabolic Networks

The known and unknown biological information in the metabolic network is visualized using a graph visualization tool. Figure 4 shows a screenshot of the FCModeler tool display window. The graph visualization is based on for visualizing and interacting with dynamic information spaces. FCModeler uses *Diva*, a Java-based software information visualization package (see http://www.gigascale.org/diva/) for its basic graph data structure, rendering, and interaction controls. In addition, it extends *Diva* to provide custom graphics-related features such as dynamic figures, graph layout, and panning and zooming. This allows for a greater variety of visualization objects on the display. The front end of the FCModeler tool is a Java ™ interface that reads and displays data from a database of links and nodes. The graph layout program is *dot*, which is part of the *Graphviz* program developed at AT&T research labs (see http://www.research.att.com/sw/tools/graphviz/).

The nodes and edges in the FCModeler graph have properties, which can be specified in an XML file or created at run-time by the user. There is a set of properties for nodes and also one for edges. In a bioinformatics application, a node property may be "type of node". Then each node would have a specific value for this property, such as "DNA", "RNA", "protein", "environmental factor", etc. Similarly, an edge property could be "type of reaction" with the specific values "conversion" or "regulatory." Figure 5 shows the visual property window from FCModeler for some of the nodes and edges of the Arabidopsis graph shown in Figure 4.

Interaction

FCModeler currently supports several forms of user interaction with the graph model and view. One basic form of interaction is selection. Node and edge figures can be selected individually by clicking on them with the mouse, or by dragging a selection rectangle around a group of them. The selected node and edge figures are then visually distinguished from the rest by some

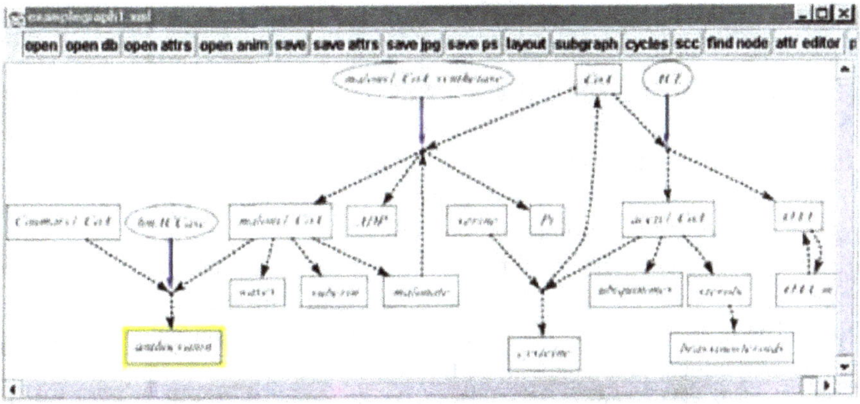

Figure 4. Screenshot of an FCModeler graph. The bold blue arrows represent catalyst links. The dashed arrows are conversion links. The proteins are shown as ellipses. The rectangles are small molecules. Nodes of interest can be highlighted by the user.

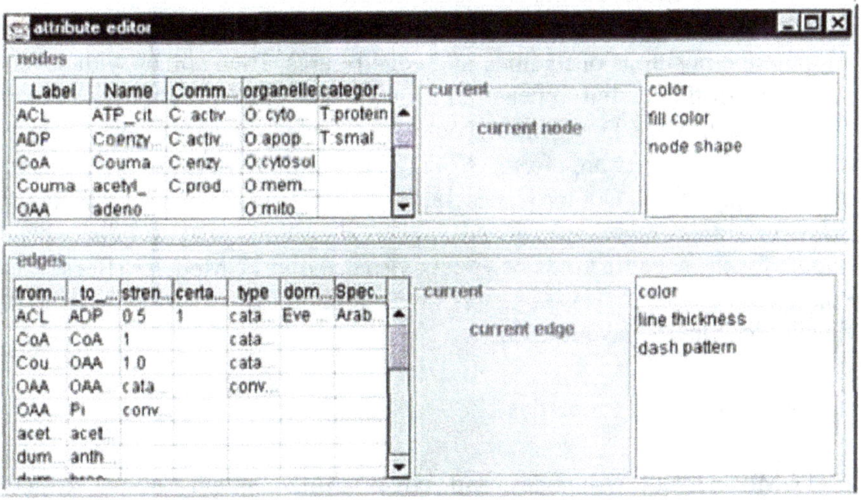

Figure 5. The attribute editor in FCModeler. The color, shape, and fill of the nodes can be changed according to the existing properties. The color, line thickness, and dash pattern can be changed for the edges.

form of highlighting. Selection of node and edge figures can provide a starting point for other operations on the graph.

The user can reposition the nodes and edges on the screen by dragging them with the mouse. All of the selected figures will then be translated in the direction of the mouse movement. In addition, edge figures are rendered as Bezier curves [Angel, 2000] and dragging with the mouse relocates the edge figures' individual control points.

FCModeler supports graphical modification of the underlying metabolic map model. Node and edge figures can be added to and removed from the view. The user can also change the tail or head node of an edge by dragging the desired edge end to a new node figure.

Zooming and panning allow the user to examine different parts of the graph in varying levels of detail. The graph may just be too large to be viewed as a whole on the screen, or a layout algorithm could use more space than is viewable at once for its layout. The view port can also be programmatically set to arbitrary coordinates.

Graph Layout

Any Diva graph view can use an arbitrary graph layout algorithm to compute the positions of its node and edge figures. Diva comes with several layout algorithms, but opens its views to custom implementations. FCModeler uses the Dot graph layout engine, which is part of the Graphviz graph drawing software from AT&T labs (http://www.research.att.com/sw/tools/graphviz/). Dot produces fairly nice layouts, and is easy to use. However, other more specialized layout algorithms may produce better layouts for the specific kinds of graphs visualized in FCModeler [Becker and Rojas, 2001]. Diva makes pluggable layout algorithms easy by separating the view logic from the layout logic.

Database and Object Properties

FCModeler allows nodes and edges in the graph model to have properties. The specific values of these properties determine the visual attributes of the corresponding node and edge figures in the view. These mappings from properties to visual attributes are encapsulated by a set of mapping rules, which can be specified in an XML file or created at run-time by the user.

The node and link information is stored in a relational database that interacts with the graphical modeling program. The purpose of this database is to store information such as links and nodes data, search results, literature sources, and microarray data in a searchable database to support development of the Gene Expression Toolkit. This system will be used to model the structure of metabolic networks using data provided by users. It will also track the results from the tests. Figure 6 shows a property window that displays the database information about the highlighted nodes and links.

Animation

The visual attributes of the node and edge figures can be changed over time, producing an animation of the graph view. This animation consists of discrete time steps, each having a set of mapping rules. An animation controller in FCModeler applies the mapping rules to the node and edge figures for each time step in order, with a configurable delay between time steps. The node and edge figures are set back to a permanent state at the beginning of each time step, and then the new mapping rules are applied to all figures in the view. Thus, the mappings only last for a single time step, and then the figures revert back to their previous state. The user specifies the sets of mapping rules for each time step of the animation in an XML file. This file is similar to the attributes XML file, but with the addition of time step tags. Users can produce these animation files to show how the nodes interact with each other in the graph.

Metabolic Network Modeling using Fuzzy Cognitive Maps

The FCModeler tool models regulatory networks so that important relationships and hypotheses can be mined from the data. Some types of models that have been studied for representing gene regulatory networks are Boolean networks [Liang *et al.*, 1998; Akutsu *et al.*, 1999], linear weighting networks [Weaver *et al.*, 1999], differential equations [Tomita *et al.*, 1999; Akutsu, 2000] and Petri nets [Matsuno, 2000]. Circuit simulations and differential equations such as those used in the E-cell project require detailed information that is not yet known about the regulatory mechanisms between genes. Another problem is the numerical instability inherent in solving large networks of differential equations. Boolean networks analyze binary state

node	Label	Name	Comments	organelle	category of compund
serine	serine	serine		O cytosol	T small molecule
cysteine	cysteine	cysteine		O cytosol	T small molecule
acetyl_CoA	acetyl_CoA	acetyl_CoA	C activated acetate in cytosol	O cytosol	T small molecule
CoA	CoA	CoenzymeA	C activates acetate, malate, fatty acids, etc	O cytosol	T small molecule

edge from_node	to_node_2	certainty	type	affected by	c. domain expert	location	Species
e467 dummy62	cysteine	1	converted		Eve Wurtele		Arabidopsis
e466 serine	dummy62	1	converted		Eve Wurtele		Arabidopsis
e468 dummy62	CoA	1	converted		Eve Wurtele		Arabidopsis
e465 acetyl_CoA	dummy62	1	converted		Eve Wurtele		Arabidopsis

Figure 6. The property viewer displays information about the selected nodes and edges. The properties are defined in an XML graph file generated by the relational database.

transition matrices to look for patterns in gene expression. Each part of the network is either on or off depending on whether a signal is above or below a pre-determined threshold. These network models lack feedback. Linear weighting networks have the advantage of simplicity since they use simple weight matrices to additively combine the contributions of different regulatory elements. However, the Boolean and weighting networks are feedforward systems that cannot model the feedback present in metabolic pathways. Petri nets can handle a wide variety of information; however their complexity does not scale up well to systems that have both continuous and discrete inputs [Alla and David, 1998; Reisig and Rozenberg, 1998].

Fuzzy cognitive maps (FCMs) have the potential to answer many of the concerns that arise from the existing models. Fuzzy logic allows a concept or gene expression to occur to a degree—it does not have to be either on or off [Kosko, 1986a]. FCMs have been successfully applied to systems that have uncertain and incomplete models that cannot be expressed compactly or conveniently in equations. Some examples are modeling human psychology [Hagiwara, 1992], and on-line fault diagnosis at power plants [Lee *et al.*, 1996]. All of these problems have some common features. The first is the lack of quantitative information on how different variables interact. The second is that the direction of causality is at least partly known and can be articulated by a domain expert. The third is that they link concepts from different domains together using arrows of causality. These features are shared by the problem of modeling the signal transduction and gene regulatory networks.

We use a series of +/– links to model known and hypothesized signal transduction pathways. Another link type suggests a relationship between concepts with no implied causality. These links will be constructed by mining the literature using PathBinder and from Gene Expression Toolkit Database that contains the expert knowledge of biologists. Given the metabolic network, FCModeler contains advanced tools that:

- Locate and visualize cycles and strongly connected components of the graph.
- Simulate intervention in the network (e.g. what happens when a node is shut off) and search for critical paths and control points in the network.
- Capture information about how edges between graph nodes change when different regulatory factors are present.

Metabolic Network Modeling

Fuzzy cognitive maps are fuzzy digraphs that model causal flow between concepts or, in this case, genes, proteins, and transcription factors [Kosko, 1986a; Kosko, 1986b]. The concepts are linked by edges that show the degree to which the concepts depend on each other. FCMs can be binary state systems called simple FCMs with causality directions that are +1, a positive causal connection, -1, a negative connection, or zero, no causal connection. The fuzzy structure allows the gene or protein levels to be expressed in the continuous range $[0,1]$. The input is the sum of the product of the fuzzy edge values. The system nonlinearly transforms the weighted input to each node using a threshold function or other nonlinear activation. FCMs are signed digraphs with feedback. Nodes stand for causal fuzzy sets where events occur to some degree. Edges stand for causal flow. The sign of an edge (+ or −) shows causal increase or decrease between nodes. The edges between nodes can also be time dependent functions that create a complex dynamical system. Neural learning laws and expert heuristics encode limit cycles and causal patterns. One learning method is differential Hebbian learning in which the edge matrix updates when a causal change occurs at the input [Dickerson and Kosko, 1994].

Each causal node $C_i(t)$ is a nonlinear function that maps the output activation into a fuzzy membership degree in $[0,1]$. Simple or trivalent FCMs have causal edge weights in the set {-1,0,1} and concept values in {0,1} or {-1,1}. Simple FCMs give a quick approximation to an expert's causal knowledge. More detailed graphs can replace this link with a time-dependent and/or nonlinear function.

FCMs recall as the FCM dynamical system equilibrates. Simple FCM inference is matrix-vector multiplication followed by thresholding. State vectors C_n cycle through the FCM edge matrix E, which defines the edges e_{ki} where k is the upstream node and i is the downstream node. The system nonlinearly transforms the weighted input to each node C_i:

$$C_i(t_{n+1}) = S\left[\sum e_{ki}(t_n)C_k(t_n)\right]$$

$S(y)$ is a monotonic signal bounded function such as the sigmoid function:

$$S_j(y_j) = \frac{1}{1+e^{-c\left(y_j - T_j\right)}}$$

In this case $c=1000$ and $T_j= 0.5$ for all nodes. This is equivalent to a step function with a threshold at 0.5. The edges between nodes can also be time dependent functions that create a complex dynamical system.

Regulatory Links: The regulatory edges are modeled using a simple FCM model that assumes binary connecting edges: $e_{ki} = \{-1,1\}$ for the single edge case. When there are multiple excitatory or inhibitory connections, the weights are divided by the number of input connections in the absence of other information. As more information becomes known about details of the regulation, for example how RNA level affects the translation of the corresponding protein, the function of the link models will be updated. The regulatory nodes will also have self-feedback since the nodes stay on until they have been inhibited.

Conversion Links: Conversion relationships are modeled in different ways depending on the goal of the simulation study. The first case corresponds to investigating causal relationships between nodes. The node is modeled in the same manner as a regulatory link in which the presence of one node causes presence at the next node. When information about the rate of change in a reaction is available, a simple difference equation can model the gradually rising and falling levels of the nodes. When stoichiometric information is available, the links can be modeled as a set of mass-balance equations. The step size depends on the reaction rate and the stoichiometric relationship between the nodes.

Catalyzed Links: Catalyzed reactions add a dummy node that acts upon a conversion link. This allows one link to modify another link. In the current model, the catalyzed link is simulated by weighting the inputs into the dummy node in such a way that both inputs must be present for the node to be active. Another method of modeling catalyzed links is an augmented matrix that operates on the edges between the nodes. The catalyst node acts as a switch that allows a reaction to occur when the proper substrates are available. Since all of the compounds must be present in these links for a reaction to occur the pieces must be modeled as a logical AND operation. This operation is commonly modeled as a minimum function; however, it can also be modeled as a product of all the input values [Kosko, 1992].

Forcing functions: In biological systems such as cells, many of the metabolic network elements are always present. This is modeled as a node is active unless it is being inhibited:

$$C_i\left(t_{n+1}\right)=S\left[\sum e_{ki}\left(t_n\right)C_k\left(t_n\right)+1\right]$$

9.6 Example of PathBinder-FCModeler Integration

This example shows how the pieces of the Gene Expression Toolkit can be used to create or update metabolic maps of a system using expert knowledge. The process starts with a map created by an expert or an existing metabolic pathway from a database such as KEGG or WIT [Kanehisa and Goto, 2000; Overbeek *et al.*, 2000]. The next step is to perform a PathBinder literature search for new relationships between the nodes of the existing graph. These relationships can then be assessed and added into the metabolic map. FCModeler models the effects of the changes for biologist user. An expert in the area of gibberellin metabolism constructed the map shown in Figure 7. Next a PathBinder Query is performed as shown below.

Query: Find sentences containing (either gibberellin, gibberellins, or GA) AND (either SPY, SPY-4, SPY-5, or SPY-7).

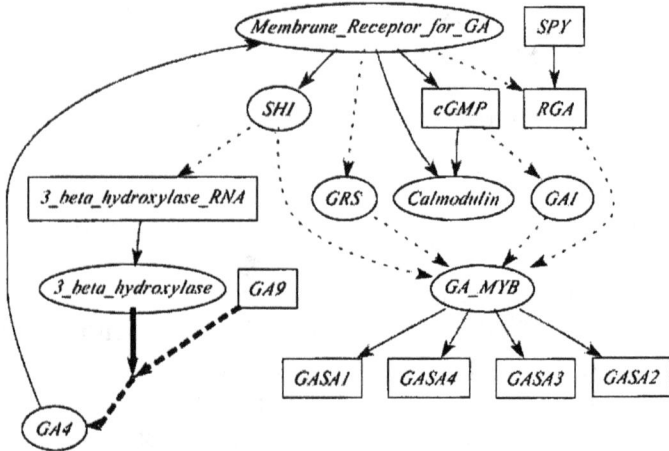

Figure 7. Hypothetical network of gibberellin metabolism and regulation in Arabidopsis. Heavy lines are catalyzed links, heavy dashed lines are conversion links, and thin lines are regulatory links. All proteins are shown in elliptical boxes.

Sentence: "Here we describe detailed studies of the effects of two of these suppressors, spy-7 and gar2-1, on several different GA-responsive growth processes (seed germination, vegetative growth, stem elongation, chlorophyll accumulation, and flowering) and on the in plant amounts of active and inactive GA species." Source: UI—99214450 Peng J, Richards DE, Moritz T, Cano-Delgado A, Harberd NP, Plant Physiol 1999 Apr;119(4): 1199-1208. Figure 8 shows the new graph after the information provided by the new links is added into the graph.

9.7 Example of Network Modeling

The metabolism and signal transduction of the plant hormone gibberellin in Arabidopsis [Hedden and Phillips, 2000; Sun, 2000] was used to test this modeling scheme. Figure 7 shows the nodes used in this test. An expert

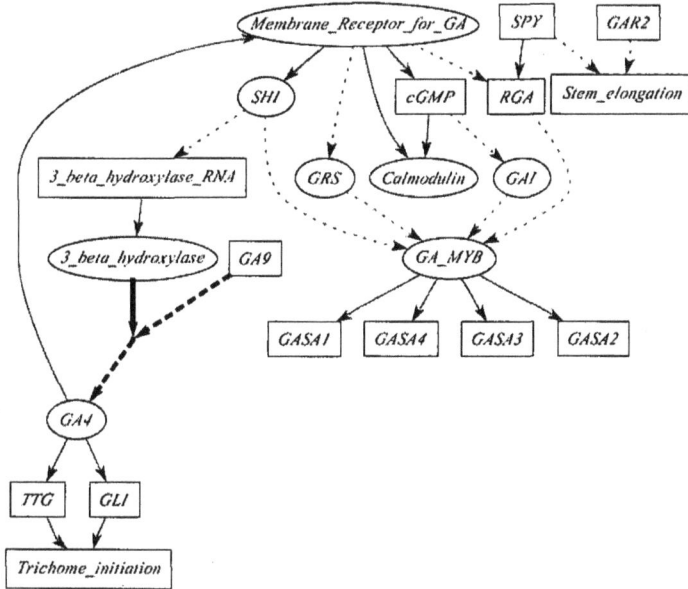

Figure 8. The updated map based on the PathBinder query result. The new nodes are shaded.

researcher in the field created the link types and causality directions. The key element in this graph is the block labeled GA4. This compound regulates many other regulatory mechanisms in plants. GAI, GRS, SPY, and GA_MYB had forcing functions applied to them. Figures 9 and 10 show visualized networks at different time steps to analyze the interactions in the network. Figure 9 shows the operation of the catalyzing node, 3_beta_hydroxylase. When the node is active, GA4 is produced. These figures show how GA4 can regulate its own production through the transcription factor SHI. The result is a homeostatic control of GA4 levels. The oscillation of the GA levels directs the generation of biomolecules that, in the absence of other constraining factors, are implicated in the formation of new cellular proliferation centers, referred to as meristems. Many key features of this model, including timing, can be tested experimentally and relatively rapidly by globally monitoring temporal profiles of mRNA, protein, and metabolite.

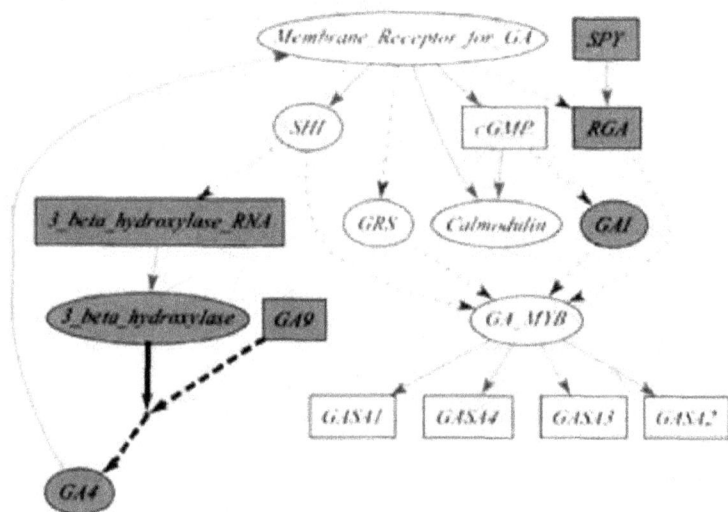

Figure 9. The catalyst, 3-beta-hydroxylase is present at this step. This allows GA9 to be converted into the active form of gibberellin, GA4. Active nodes are shaded. The nodes, SPY, GRS, and GAI are forced high in this simulation.

9.8 Conclusions

The integration of a graph visualization tool with literature mining and directed searches in microarray data allows biologists to gather and combine information from the literature, their expert knowledge, and the public databases of mRNA results. Metabolic and regulatory networks can be modeled using fuzzy cognitive maps. Future plans include: simulating intervention in the network (e.g. what happens when a node is shut off), searching for critical paths and control points in the network, and capturing information about how edges between graph nodes change when different regulatory factors are present.

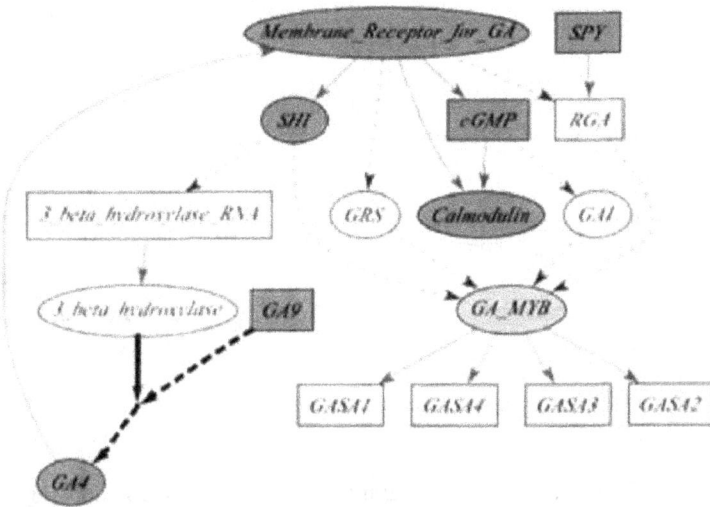

Figure 10. GA4 regulates its own production in part through thê putative DNA regulatory factor SHI. SHI inhibits the 3-beta-hydroxylase-RNA, which eventually shuts down the production of GA4.

Acknowledgments

This work is supported by grants from Proctor and Gamble Corporation, the NSF (MCB-9998292), and the Plant Sciences Institute at Iowa State University.

References

Akutsu, T., Miyano, S. and Kuhara, S. (1999) "Identification of genetic networks from a small number of gene expression patterns under the boolean network model." *Pacific Symposium on Biocomputing* **4**, 17-28.

Akutsu, T., Miyano, S. and Kuhara, S. (2000) "Algorithms for inferring qualitative models of biological networks." *Pacific Symposium on Biocomputing* **5**, 290-301.

Alla, H. and David, R. (1998) "Continuous and hybrid petri nets." *Journal of Circuits, Systems, and Computers* **8**, 159-188.

Andrade, M.A. and Valencia, A. (1998) "Automatic extraction of keywords from scientific text: Application to the knowledge domain of protein families." *Bioinformatics* **14**, 600-607.

Angel, E. (2000) *Interactive Computer Graphics.* Addison Wesley Longman, Massachusetts.

Angeline, P.J. (1996) *Advances in Genetic Programming II.* MIT Press, Cambridge, Massachusetts.

Becker, M. and Rojas, I. (2001) "A graph layout algorithm for drawing metabolic pathways." *Bioinformatics* **17**, 461-467.

Berleant, D. (1995) "Engineering word experts for word disambiguation." *Natural Language Engineering* **1**, 339-362.

Blaschke, C., Andrade, M., Ouzounis, C. and Valencia, A. (1999) "Automatic extraction of biological information from scientific text: Protein-protein interactions." *International Conference on Intelligent Systems for Molecular Biology,* Heidelberg.

Brown, M.P.S., Grundy, W.N., Lin, D., Cristianini, N., Sugnet, C.W., Furey, T.S., Ares, M. and Haussler, D. (2000) "Knowledge-based analysis of microarray gene expression data by using support vector machines." *Proceedings National Academy of Science* **97**, 262-267.

Collier, N.H., Park, S., Ogata, N., Tateishi, U.Y., Nobata, C., Ohta, T., Sekimizu, T., Imai, H., Ibushi, K. and Tsujuii, J. (1999) "The GENIA project: Corpus-based knowledge acquisition and information extraction from genome research papers." *European Association for Computational Linguistics (EACL) Conference.*

Craven, M. and Kumlien, J. (1999) "Constructing biological knowledge bases by extracting information from text sources." *AAAI Conference on Intelligent Systems in Molecular Biology,* 77-86.

Dickerson, J.A. and Kosko, B. (1994) "Virtual worlds as fuzzy cognitive maps." *Presence* **3**, 173-189.

Ding, J., Berleant, D., Nettleton, D. and Wurtele, E. (2002) "Mining MEDLINE: Abstracts, sentences, or phrases?" *Pacific Symposium on Biocomputing*, Kaua'i, Hawaii.

Eisen, M.B., Spellman, P.T., Brown, P.O. and Botstein, D. (1998) "Cluster analysis and display of genome-wide expression patterns." *Proceedings National Academy of Science* **95**, 14863-14868.

Fukuda, K., Tsunoda, T., Tamura, A. and Takagi, T. (1998) "Toward information extraction: Identifying protein names from biological papers." *Pacific Symposium on Biocomputing*, 707-718.

Hagiwara, M. (1992) "Extended fuzzy cognitive maps." *IEEE Int Conf Fuzzy Syst FUZZ-IEEE*, 795-801.

Hedden, P. and Phillips, A.L. (2000) "Gibberellin metabolism: New insights revealed by the genes." *Trends Plant Sci.* **5**, 523-530.

Humphreys, K., Demetriou, G. and Gaizauskas, R. (2000) "Two applications of information extraction to biological science journal articles: Enzyme interactions and protein structures." *Pacific Symposium on Biocomputing* **5**, 502-513.

Kanehisa, M. and Goto, S. (2000) "KEGG: Kyoto encyclopedia of genes and genomes." *Nucleic Acids Research* **28**, 27-30.

Karp, P.D., Krummenacker, M., Paley, S. and Wagg, J. (1999) "Integrated pathway/genome databases and their role in drug discovery." *Trends in Biotechnology* **17**, 275-281.

Karp, P.D., Riley, M., Saier, M., Paulsen, I.T., Paley, S.M. and Pellegrini-Toole, A. (2000) "The EcoCyc and MetaCyc databases." *Nucleic Acids Research* **28**, 56-59.

Kinnear, K.E. (1994) *Advances in Genetic Programming*. MIT Press, Cambridge, Massachusetts.

Kosko, B. (1986a) "Fuzzy cognitive maps." *International Journal of Man-Machine Studies* **24**, 65-75.

Kosko, B. (1986b) "Fuzzy knowledge combination." *International Journal of Intelligent Systems* **1**, 293-320.

Kosko, B. (1992) *Neural Networks and Fuzzy Systems*. Prentice Hall, Englewood Cliffs, New Jersey.

Koza, J.R. (1992) *Genetic Programming: On the Programming of Computers by Means of Natural Selection.* MIT Press, Cambridge, Massachusetts.

Koza, J.R. (1994) *Genetic Programming II: Automatic Discovery of Reusable Programs.* MIT Press, Cambridge, Massachusetts.

Lee, K., Kim, S. and Sakawa, M. (1996) "On-line fault diagnosis by using fuzzy cognitive maps." *IEICE Transactions on Fundamentals of Electronics, Communications and Computer Sciences* **E79-A**, 921-922.

Liang, S., Fuhrman, S. and Somogyi, R. (1998) "REVEAL, a general reverse engineering algorithm for inference of genetic network architectures." *Pacific Symposium on Biocomputing* **3**, 18-29.

Matsuno, H., Doi, A., Nagasaki, M. and Miyano, S. (2000) "Hybrid petri net representation of gene regulatory network." *Pacific Symposium on Biocomputing* **5**, 338-349.

Ng, S.K. and Wong, M. (1999) "Toward routine automatic pathway discovery from on-line scientific text abstracts." *Genome Informatics* **10**, 104-112.

Ono, T., Hishigaki, H., Tanigami, A. and Takagi, T. (2001) "Automated extraction of information on protein-protein interaction from the biological literature." *Bioinformatics* **17**, 155-161.

Overbeek, R., Larsen, N., Pusch, G.D., D'Souza, M., Selkov, E. Jr., Kyrpides, N., Fonstein, M., Maltsev, N. and Selkov, E. (2000) "WIT: Integrated system for high-throughput genome sequence analysis and metabolic reconstruction." *Nucl. Acids. Res.* **28**, 123-125.

Proux, D., Rechenmann, F., Julliard, L., Pillet, V. and Jacq, B. (1998) "Detecting gene symbols and names in biological texts: A first step toward pertinent information." *Ninth Workshop on Genome Informatics,* 72-80.

Reisig, W. and Rozenberg, G. (1998) *Lectures on Petri Nets I: Basic Models.* Springer, Berlin.

Rindflesch, T.C., Hunter, L. and Aronson, A.R. (1999) "Mining molecular binding terminology from biological text." *American Medical Informatics Association (AMIA) '99 Annual Symposium,* 127-131.

Rindflesch, T.C., Tanabe, L., Weinstein, J.N. and Hunter, L. (2000) "EDGAR: Extraction of drugs, genes, and relations from the biomedical literature." *Pacific Symposium on Biocomputing,* 514-525.

Sekimizu, T., Park, H.S. and Tsujii, T. (1998) *Genome Informatics.* Universal Academy Press.

Shatkay, H., Edwards, S., Wilbur, W.J. and Boguski, M. (2000) "Genes, themes and microarrays: using information retrieval for large-scale gene analysis." *8th International Conference on Intelligent Systems for Molecular Biology,* 317-328.

Stapley, B.J. and Benoit, G. (2000) "Biobibliometrics: Information retrieval and visualization from co-occurrences of gene names in MEDLINE abstracts." *Pacific Symposium on Biocomputing* **5**, 529-540.

Sun, T. (2000) "Gibberellin signal transduction." *Curr. Opin. Plant Biol.* **3**, 374-380.

Tanabe, L., Scherf, U., Smith, L.H., Lee, J.K., Hunter, L. and Weinstein, J.N. (1999) "MedMiner: An internet text-mining tool for biomedical information, with application to gene expression profiling." *BioTechniques* **27**, 1210-1217.

Thomas, J., Milward, D., Ouzounis, C., Pulman, S. and Carrol, M. (2000) "Automatic extraction of protein interactions from scientific abstracts." *Pacific Symposium on Biocomputing* **5**, 538-549.

Tomita, M. (2001) "Whole-cell simulation: A grand challenge of the 21st century." *Trends Biotechnol.* **19**, 205-210.

Tomita, M., Hashimoto, K., Takahashi, K., Shimizu, T.S., Matsuzaki, Y., Miyoshi, F., Saito, K., Tanida, S., Yugi, K., Venter, J. and Hutchison, C. (1997) "E-CELL: Software environment for whole cell simulation." *Workshop on Genome Informatics,* 147-155.

Tomita, M., Hashimoto, K., Takahashi, K., Shimizu, T.S., Matsuzaki, Y., Miyoshi, F., Saito, K., Tanida, S., Yugi, K., Venter, J. C. and Hutchison, C. A. (1999) "E-CELL: Software environment for whole-cell simulation." *Bioinformatics* **15**, 72-84.

Usuzaka, S., Sim, K.L. and Tanaka, M. (1998) "A machine learning approach to reducing the work of experts in article selection from database: A case study for regulatory relations of *S. cerevisiae* genes in MEDLINE." *Ninth Workshop on Genome Informatics,* 91-101.

Weaver, D.C., Workman, C.T. and Stormo, G.D. (1999) "Modeling regulatory networks with weight matrices." *Pacific Symposium on Biocomputing* **4**, 112-123.

Wong, L. (2001) "PIES, a protein interaction extraction system." *Pacific Symposium on Biocomputing* **6**, 520-531.

Authors' Addresses

J.A. Dickerson, Electrical and Computer Engineering Department, Iowa State University, Ames, Iowa, USA. Email: julied@iastate.edu.

D. Berleant, Electrical and Computer Engineering Department, Iowa State University, Ames, Iowa, USA.

Z. Cox, Electrical and Computer Engineering Department, Iowa State University, Ames, Iowa, USA.

W. Qi, Electrical and Computer Engineering Department, Iowa State University, Ames, Iowa, USA.

D. Ashlock, Mathematics Department, Iowa State University, Ames, Iowa, USA.

E.S. Wurtele, Botany Department, Iowa State University, Ames, Iowa, USA.

A.W. Fulmer, Proctor & Gamble Corporation, Cincinnati, Ohio, USA.

Chapter 10

Phyloinformatics and Tree Networks

William H. Piel

10.1 Introduction

All organisms on Earth are related to one another through an enormous "tree of life" that had its origins some four billion years ago. This unifying aspect to life means that genomics, evolution, and development are inextricably linked, and understanding how, for example, genotype becomes phenotype, requires that we know phylogeny almost as well as we know genomics [Eisen, 1998; Mizuno *et al.*, 2001]. Indeed, the concept of homology among genes, and homology among the functional behaviors of genes, goes to the very heart of phylogeny: homology arises when a derived trait is shared among descendant species as seen on a phylogenetic tree. Many biologists believe that phylogenetics will play a crucial role in bioinformatics and functional genomics, not to mention ecology, evolution, and behavior [Pennisi, 2001].

In recent years, the number of publications that describe new phylogenies appears to have grown almost exponentially. Between 1989 and 1991, Sanderson *et al.* [1993] compiled a list of the number of trees published in as many journals as they could find phylogenetic reports. They found approximately 40% growth in just three years. Separately, the Jungle database (http://smiler.lab.nig.ac.jp/jungle/jungle.html) collected trees from twelve selected journals over periods ranging from four to twelve years. The curve of the sum of regression lines for the growth among these twelve journals approximates the slope of [Sanderson *et al.*, 1993] over the same three-year period, but the Jungle data confirmed that the trend continued for another six years (Figure 1). Judging by these trends, one could estimate that by the year 2000 probably over 1,000 publications per year would have contained phylogenies. Given this growth rate, the data generated by this

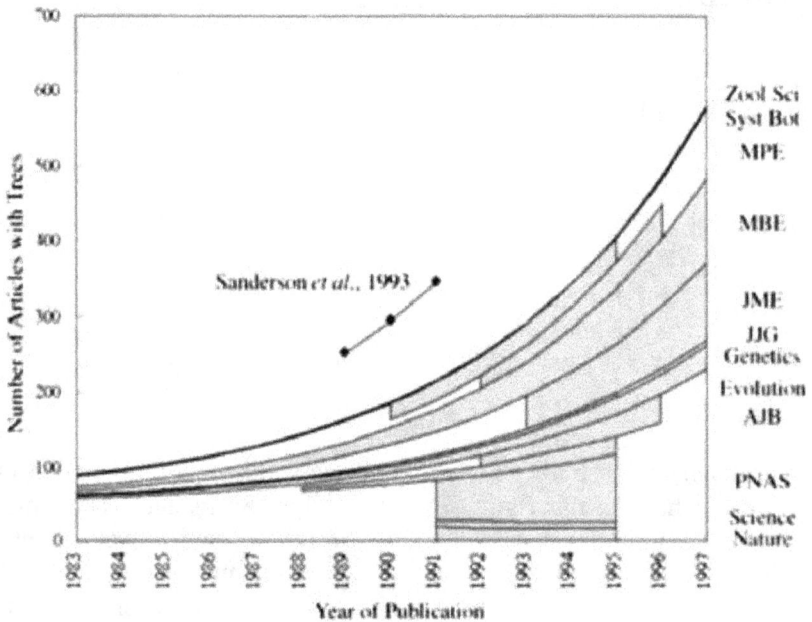

Figure 1. The growth of phylogenetic data, as compiled by [Sanderson *et al.*, 1993] and the Jungle database (Saitou *et al.*, http://smiler.lab.nig.ac.jp/jungle/jungle.html). The shaded sections of the graph indicate the time periods when the twelve selected journals were scanned for trees. For each journal a regression of its growth curve was calculated. These lines were stacked on the graph to produce a sum total. The summed growth rate among the twelve journals over the 1989-1991 period is only marginally lower than the estimates of [Sanderson *et al.*, 1993] for all journals. Therefore it is probably safe to extrapolate the [Sanderson *et al.*, 1993] data, thereby predicting that by the year 2000 over 1,000 publications per year contained phylogenies.

discipline will get out of hand unless the results are properly complied and informaticized.

Despite the growth of phylogenetic data, methods for organizing, data mining, and inferring a synthesis of phylogenetic knowledge are poorly developed. Storing phylogenies in a database is, by itself, a challenging task, as the data elements are diverse and complex. The essential information that needs to be stored is recorded in the topology of hierarchically nested nodes, the distances between them, and their identities. Consequently, the usual data searching methods, such as pattern recognition as used in sequence databases, are not useful here: a whole new conceptual approach is in order.

In 1994 work began on TreeBASE, a database of phylogenetic knowledge [Sanderson *et al.*, 1994]. The initial purpose was to develop a simple means of storing phylogenetic trees and the aligned data matrices used to produce them, if nothing else but to grasp a handle on the burgeoning growth of published works in the field. Initially, it merely resembled an enhanced, specialized literature database, in which trees and matrices were stored in addition to the usual citation and abstract. However, it soon became apparent that the user needed specialty tools for data mining and meta analysis so as to better navigate the data. To this end, "tree surfing" was implemented as a means of locating neighboring trees [Piel *et al.*, 2002a], and the concept of the "tree-graph" was described for identifying clusters of candidate trees for supertree construction [Sanderson *et al.*, 1998]. Seamless connection with a supertree server (http://darwin.zoology.gla.ac.uk/cgi-bin/supertree.pl) facilitated subsequent supertree construction.

While tools for more effective data searches are useful, ultimately the database should be designed so that an overall phylogenetic picture continuously emerges with the growth of individual phylogenetic elements. This improvement would, in effect, incorporate a computational symbolic theory into the database model — i.e., it would implement a formal ontology which is then available for computational analysis. The effect is to transform a database of phylogenetic publications, in which the burden is on the user to reexamine the data and infer the "big picture," to one that resembles an artificial intelligence, in which the burden has shifted to the computer [Karp, 2001]. Karp [2001] argues that a computational symbolic theory would allow inferences to emerge from the computer that are otherwise too complex for any one biologist to grasp, such as the web of biochemical pathways in *Escherichia coli* as is stored in the EcoCyc database.

In this chapter I illustrate how networks of trees can help locate "neighborhoods" of trees and ultimately interconnect all trees in a database.

In addition, I discuss how the structure of the database could be modified so as to make the stored phylogenies more readily available for computational analysis.

10.2 Small-World Networks

Small-world networks belong to a special class of disordered networks that has short characteristic path lengths despite appearing to be highly clustered and non-random [Watts and Strogatz, 1998; Watts, 1999]. This phenomenon is popularly known as the "six degrees of separation," in which it is thought that no two people on earth are separated by a chain of friendships that exceeds six people, despite the fact that each person's friendship clique is largely local and non-random. For epidemiologists, this notion has important ramifications for disease propagation [Watts and Strogatz, 1998], where, for example, it is thought that just a few promiscuous individuals have a disproportionately large impact on the spread of venereal disease [Liljeros *et al.*, 2001]. For molecular biologists, small-world networks of cellular proteins can help identify the most vital molecules for cell survival on the basis of high levels of connectivity [Jeong *et al.*, 2001].

Phylogenetic trees can be connected with one another by sharing the same taxonomic identities, much the way two people might be connected by having a friend in common. The web of connections among members of a collection of trees forms a network that is neither completely random nor completely regular [Piel *et al.*, 2002b]. Navigating through this web of trees is what, in TreeBASE, is called tree surfing. This tool involves searching on all taxa in a set of trees to recover a larger set of trees, *et cetera*, and with each subsequent iteration a new set of trees is found with yet one more "degree of separation" from the previous set. The collective dynamics of small-world tree networks have important implications as to how effective automated supertree tools might be at turning a neighborhood of trees into a single consensus phylogeny. Supertree methods themselves are still primitive and under development [Sanderson *et al.*, 1998; Semple and Steel, 2000], but mining a phylogenetic database to recover the best candidate trees is a critical preliminary step, and that is where tree networks play an important role.

The collective dynamics of tree networks can tell us, for example, whether the diameter of the network — as measured by the characteristic path length (L), that being the average degree of separation between any pair of trees — will expand as the database grows until surfing the network becomes too laborious. Alternatively, the diameter of the network might implode, rendering tree surfing too coarse. We can also examine progress in connecting trees so as to minimize the persistence of disconnected satellites within a database. Recent research in disordered networks has shown how these characteristics can be probed, such as by examining the distribution function of connectivities [Watts and Strogatz, 1998; Barbási and Albert, 1999; Watts, 1999; Amaral *et al.*, 2000; Strogatz, 2001; Piel *et al.*, 2002b].

10.3 Tree Networks vs. Random Networks

As compared to random networks, small-world networks are characterized by a relatively small diameter (*L*) while retaining neighborhoods that are seemingly non-random [Watts, 1999]. To examine these two characteristics in small-world networks, Watts and Strogatz [1998] used a clustering coefficient (*C*) to estimate non-randomness, defined as the fraction of possible edges in each vertex's neighborhood that actually exist, averaged over all vertices:

$$C = \left(\sum_{i=1}^{N} \frac{2v_i}{k_i(k_i - 1)} \right) \Big/ N \tag{1}$$

where k_i is the number of neighbors of the *i*th vertex, v_i is the number of edges among these neighbors, and *N* is the number of vertices.

By comparing actual networks with rewired (or permuted) networks, Watts and Strogatz [1998] showed there to be a greater relative difference for *C* than for *L* when comparing actual and randomized networks. Analysis of tree networks also showed the same effect (Table 1), albeit to a lesser degree [Piel *et al.*, 2002b]. The larger diameter of the tree network ($L = 5.11$) and the high cliquishness of trees ($C = 0.813$) probably reflects the fact that a concealed tree of life invariably shadows and shapes the topology of all separately published sub-trees. It is only the wide-ranging, deep phylogenies that join together distant neighborhoods of trees, and even this process is theoretically less random than, for example, the chaotic fluidity that characterizes distant friendship connections. But in any case, the fact that it takes over five degrees of separation on average to surf the entire tree network serves our purposes because it makes each iteration of the process more discriminating.

Network	Vertices	L_{actual}	L_{rand}	C_{actual}	C_{rand}
Film Actors[*]	225,226	3.65	2.99	0.79	0.00027
Power Grid[*]	4,941	18.7	12.4	0.08	0.005
C. elegans[*]	282	2.65	2.25	0.28	0.05
TreeBASE[†]	989	5.11	2.00	0.813	0.182

Table 1. Characteristic path length (L) and clustering coefficient (C) for four actual networks compared to permuted ones. Sources: [*]Watts, D.J. and Strogatz, S.H. (1998) "Collective dynamics of 'small-world' networks." *Nature* **394**, 440-442.

10.4 The Growth of Tree Networks

As journals publish new phylogenies, and as the systematics community gears up to tackle the entire tree of life [Pennisi, 2001], it is not obvious how a database of phylogenies will grow. In particular, to what extent will a single, well-connected island network emerge from a growing collection of trees? Since supertree algorithms can only function properly once pairs of trees in a network share at least two connections between them, it is worth exploring how many trees are needed before a supertree could possibly emerge.

If TreeBASE is a fair sample of published trees in the literature, it would appear that collections of trees begin assembling into a single, large island (i.e., a dominant grouping of interconnected trees) even after only 250 trees [Piel *et al.*, 2002b]. Random subsets of the database show an initial drop in the size of the largest island as a percentage of all trees, indicating that when new trees are added to a database of less than 250 trees, they are more likely to be disconnected than connected (Figure 2). However, between 500 and 700 trees, the largest island jumps from 20-30% to 60-70% of all trees, equivalent to the sudden coalescence among components of a random graph [Erdös and Rényi, 1960]. By 1,200 trees, the largest island holds almost 80%

of all trees (Figure 2), and will, in theory, eventually reach 100%. Similarly, the growth in the absolute number of islands indicates that after about 1,200 trees it reaches a stable point of about 100 islands (Figure 2, solid diamonds), and these will, presumably, decrease in number as new trees succeed in bridging them to the main island. This stable point could be called "island parity," wherein new trees no longer cause a net growth in the number of islands.

Although island parity first occurs with surprisingly few trees in the database, the definition of neighbor used here requires only one taxon shared between two trees in order to achieve a connection. Seeing that supertree algorithms usually require more than one taxon in common, it is worth considering how more stringent definitions of neighbor affect the emergence of island parity. The number of islands plotted against database size for more stringent definitions is also shown in Figure 2 (squares, triangles, and circles). None of these have achieved island parity with the 1,300 trees available in TreeBASE; however all curves closely match second order polymorphic regressions ($R^2 = 0.99$). These regressions predict that parity will occur at about 1,700, 2,300, and 3,000 trees for stringencies of two, three, and four taxa respectively. These predictions appear to follow a simple linear function, where parity happens after $600S + 550$ trees, given stringency S. The bottom line is that a database of trees would seem to agglomerate into an interconnected network without needing to be excessively large, even with stringent definitions of neighbor.

10.5 The Distribution Function of Tree Networks

Examination of the cumulative distribution function — a curve depicting the running sum of the probabilities or frequencies of connecting to k vertices plotted against k — helps to identify idiosyncrasies and tendencies particular to a large network [Barbási and Albert, 1999; Amaral *et al.*, 2000]. The so-called scale-free power-law distribution comes about when new vertices connect preferentially to the more popular pre-existing vertices [Barbási and Albert, 1999]. Other types of distribution functions include single-scale networks with fast decaying tails, and broad-scale networks with sharp cut-offs to their power law regimes. These functions can be shaped by such factors as the aging of vertices, as seen when members of the film actors network retire. Or, for example, when the cost of adding new edges rises with the popularity of vertices, such as when airports in the airline network reach

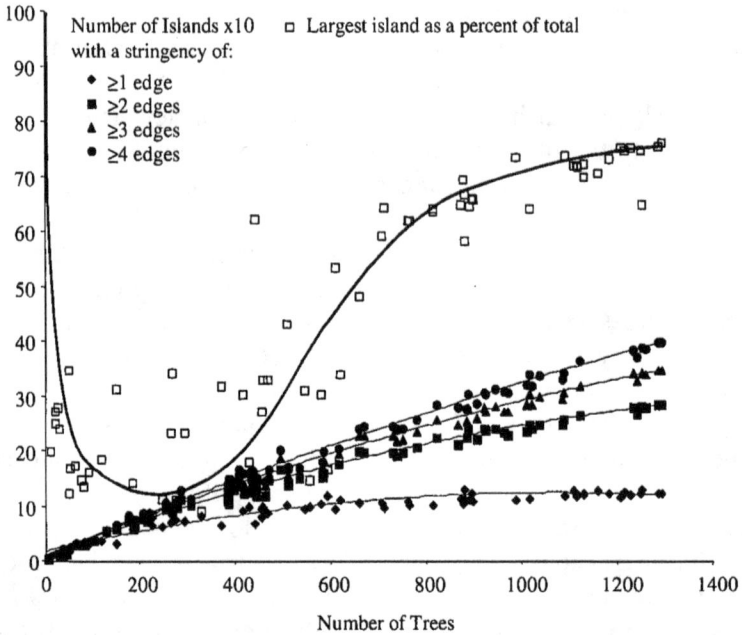

Figure 2. The growth of islands among trees in TreeBASE as a function of database size. Open squares (□) indicate the size of the largest island as a percent of database size. Closed markers indicate a tenth of the number of islands in the database given different levels of stringency for the definition of neighbor. Databases of different sizes were created by randomly selecting studies in TreeBASE and building subsets of trees based on each selection of studies [Piel *et al.*, 2002b]. Markers: diamond (♦) stringency of ≥ 1 edges between vertices, regression $y = -1\text{E-04}x^2 + 0.244x$, $R^2 = 0.92$; square (■), stringency of ≥ 2 edges, $y = -1\text{E-04}x^2 + 0.394x$, $R^2 = 0.99$; triangle (▲), stringency of ≥ 3 edges, $y = -8\text{E-05}x^2 + 0.371x$, $R^2 = 0.99$; circle (●), stringency of ≥ 4 edges, $y = -6\text{E-05}x^2 + 0.383x$, $R^2 = 0.99$.

capacity [Amaral *et al.*, 2000]. While trees in tree networks do not age as such, it is likely that taxa commonly encountered in trees will become ever more popular since more of their sequence data become publicly available. It is expected that various artifacts and idiosyncrasies associated with phylogenetic work, and with the collection of phylogenetic results, will cause unusual effects on the distribution function of tree networks.

Indeed, tree networks seem to have a distribution function unlike any class of network reported thus far [Amaral *et al.*, 2000]. This function could be called a dual-scale power law regime, in which there is a sudden change in the log-log function among trees with more than 25 neighbors (Figure 3, lines A and A'), not unlike the elbow joint in an arm [Piel *et al.*, 2002b]. The probability distribution exponent of the first part is 1.9 while the second is 4.8. These values compare with 2.3 for the actors network, 2.1 for the WWW network, and 4 for the US power grid [Barbási and Albert, 1999]. While the cause of this unusual curve is not known, excluding redundant most parsimonious trees that are largely superfluous helps to straighten the function (Figure 3, lines B and B'), but fails to eliminate it entirely [Piel *et al.*, 2002b]. The effect of this correction confirms that artifacts in the way in which biologists produce and collect trees might be responsible for the

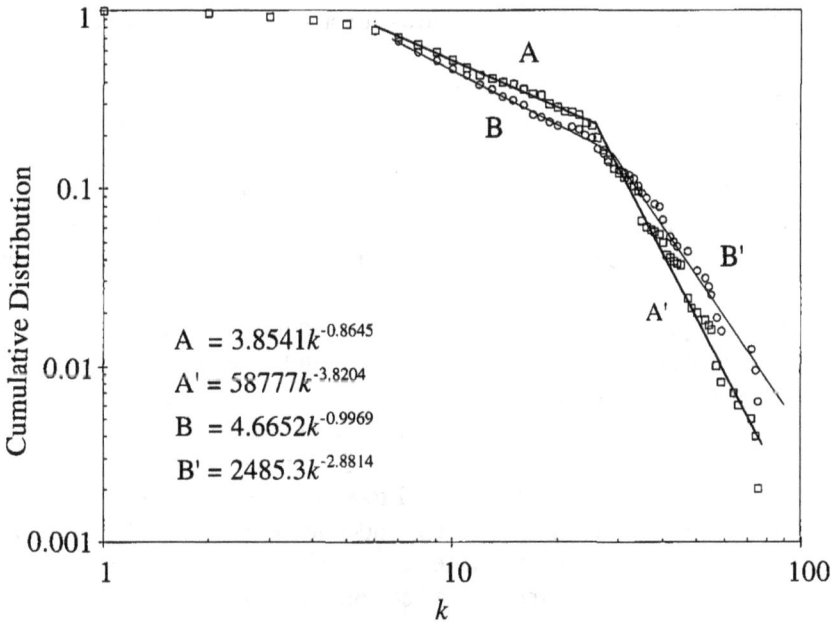

Figure 3. The distribution function of connectivities for the main island in TreeBASE. Line A and A' represents the cumulative sum of the frequencies with which each tree connects to k other trees for an island of 989 vertices with an average connectivity $\langle k \rangle = 14.99$. Line B and B' represents the same island after excluding redundant most parsimonious trees that are otherwise linked to identical sets of taxa.

curious network dynamics. This unusual distribution function might imply that different supertree strategies should be applied to different classes of more or less densely packed neighborhoods as revealed by the distribution function.

10.6 Future Developments with Tree Networks

Ultimately, networks of trees should allow databases to generate three-dimensional graphs where a diffuse shape of points takes on the vague appearance of the tree of life — each point being a tree, where tree-to-tree distances are a function of degrees of separation under various levels of stringency. Supertree algorithms can then go to work on different parts of these clouds, trying to whittle them down to a single, common consensus phylogeny.

However, with this approach the database model does not take the topology of trees into consideration. Instead, it can only sort out the patterns of overlapping sets of taxa among neighborhoods of trees. Perhaps this approach is not sufficiently sensitive to differences between trees, instead treating each tree as a single unit. It would be better if trees were built into the database model so that details of their topologies participate in assembling tree-graphs and the like.

Published phylogenies usually appear in a graphical form that reports the hierarchy of nested clades, the names of taxa, and possibly the amount of support for clades, branch lengths, or the names of clades (Figure 4A). In digital form, trees are usually stored using Newick notation, where nested parentheses correspond to nested clades, and commas represent branches (Figure 4B). TreeBASE uses this method to store trees because it takes up very little disk space and it is readily understood by most phylogenetic analysis software. One disadvantage of Newick is that trees do not incorporate into the database model, and so calculations cannot be performed directly on the trees without first decoding them.

An alternative to Newick is to disassemble each tree into its component nodes, and then store each child node and the identity of its parent node as records in a table (Figure 4C). With this method, each node receives a unique node ID number that can then identify both parent and child. Like links in a chain, the tree can be reassembled with an algorithm that recursively works it way along every branch in the tree, looking up database records of nodes as it goes. Although retrieving trees may be slower than with Newick notation,

A

E
B C
1579 1581 D
A 1582
1578
1580
1577 92%
83%
1576

B

((A,B),(C,D)E)

C

Records	1	2	3	4	5	6	7
Node ID	1576	1577	1578	1579	1580	1581	1582
Parent		1576	1577	1577	1576	1580	1580
Label			A	B	E	C	D
Taxon ID			3453	4563	8634	1297	1533
Order		1	1	2	2	1	2
Bootstrap		83			92		
Branch Lengths		2	3	5	3	4	3

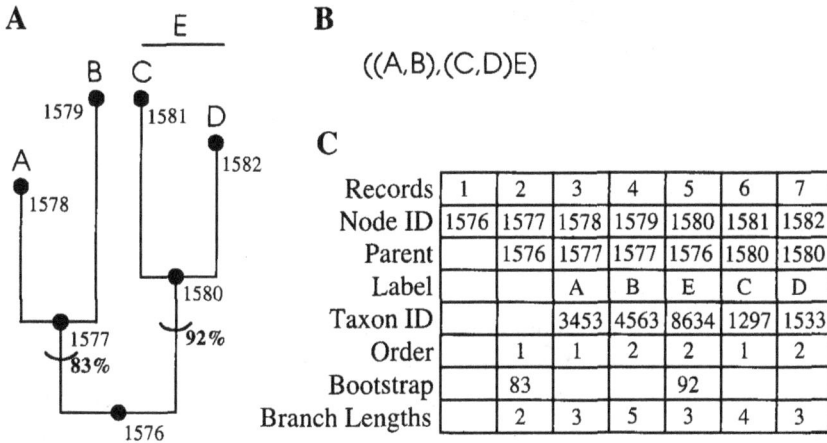

Figure 4. Parenthetical notation and parent-child records as alternative methods of storing trees. Trees in the published literature frequently appear as in 4A, a hierarchical series of nested clades. A tree's leaves are named (taxa A, B, C, and D), as are internal nodes in some instances (taxon E). The lengths of branches are sometimes represented, as are measures of branch support, such as the 83% and 92% bootstrap support exemplified here. This tree can be represented in so-called Newick notation or parenthetical notation (4B). Alternatively, the nodes of a tree can be stored as separate parent-child records in a database (4C).

this parent-child method can store lots of auxiliary information about each node (such as branch length, bootstrap, or Bremer support), which is harder to do when using Newick.

Moreover, storing each node separately means that the database has access to nested sets of clades, and therefore ultimately to the topologies of stored trees. The challenge is to make use of this added information in the process of building networks of trees. For example, instead of having vertices represent trees and edges represent taxa, the network could have the reverse arrangement, where distances between taxa are judged by degree of separation via the clades or nodes that make up trees. Taxa will cluster in taxon-graphs due to the tangled web of interconnecting nodes that attract them together, and selecting these taxa has the effect of selecting an appropriate collection of trees for subsequent supertree construction.

The MinCutSupertree algorithm for building supertrees blends many trees together into a single network of taxa, and then recursively cuts away the taxa that stem from the roots of the original trees [Semple and Steel, 2000]. With each step, the trees shorten from their bases, the network of remaining taxa shrinks, and the taxa that are cut away attach themselves to an ever emerging supertree. The parent-child method of storing phylogenetic information lends itself well to this general approach of supertree construction, in the sense that the database will have already built a giant network among all taxa.

Somewhat similar to MinCutSupertrees, reconciliation of host/parasite trees can be achieved using a network of possible solutions (a Jungle) connecting the nodes in two conflicting trees [Charleston, 1998; Page and Charleston, 1998]. After weighing possible solutions (e.g., host switching vs. lineage sorting, etc.), dynamic programming finds the shortest path through the network, and hence the most optimal solution. A MiniCutSupertree method can use a similar approach to resolve among conflicting trees, by weighing possible solutions using bootstrap or Bremer support values stored with child node records in TreeBASE.

Ideally, we would want a phylogenetic database that could store the results of calculations among existing trees in the database such that when new trees are added only a smaller subset of these calculations need to be modified to accommodate the new tree. Making use of previously stored calculations in this giant network among taxa might greatly accelerate the process of building a supertree, not unlike the way dynamic programming avoids having to recalculate those pathways in a network that it has already encountered. The bottom line is that phyloinformatic databases need to move beyond acting as highly specialized literature databases, and instead take on the role of actively building a synthesis of accumulated phylogenetic knowledge.

References

Amaral, L. A. N., Scala, A., Barthélémey, M. and Stanley, H. E. (2000) "Classes of small-world networks." *Proc. Natl. Acad. Sci. USA* **97**, 11149-11152.

Barbási, A.L. and Albert, R. (1999) "Emergence of scaling in random networks." *Science* **286**, 509-512.

Charleston, M. A. (1998) "Jungles: A new solution to the host/parasite phylogeny reconciliation problem." *Math. Biosci.* **149**, 191-223.

Eisen, J. A. (1998) "Phylogenomics: Improving functional predictions for uncharacterized genes by evolutionary analysis." *Genome Res.* **8**, 163-167.

Erdös, P. and Rényi, A. (1960) "On the evolution of random graphs." *Publ. Math. Inst. Hung. Acad. Sci.* **5**, 17-61.

Jeong, H., Mason, S. P., Barabási, A.L. and Oltvai, Z. N. (2001) "Lethality and centrality in protein networks." *Nature* **411**, 41-42.

Karp, P. (2001) "Pathway databases: A case study in computational symbolic theories." *Science* **293**, 2040-2044.

Liljeros, F., Edling, C. R., Amaral, L. A. N., Stanley, H. E. and Åberg, Y. (2001) "The web of human sexual contacts." *Nature* **411**, 907-908.

Mizuno, H., Tanaka, Y., Nakai, K. and Sarai, A. (2001) "ORI-GENE: Gene classification based on the evolutionary tree." *Bioinformatics* **17**, 167-173.

Page, R. D. M. and Charleston, M. A. (1998) "Trees within trees: Phylogeny and historical associations." *TREE* **13**, 356-359.

Pennisi, E. (2001) "Preparing the ground for a modern 'tree of life'." *Science* **293**, 1979-1980.

Piel, W. H., Donoghue, M. J. and Sanderson, M. J. (2002a) "TreeBASE: A database of phylogenetic information." In: *To the Interoperable Catalog of Life with Partners, Species 2000 Asia Oceania.* Tsukuba, Japan.

Piel, W. H., Sanderson, M. J. and Donoghue, M. J. (2002b) "The small-world dynamics of tree networks and data mining in phyloinformatics." in prep.

Sanderson, M. J., Baldwin, B. G., Bharathan, G., Campbell, C. S., Von Dohlen, C., Ferguson, D., Porter, J. M., Wojciechowski, M. F. and Donoghue, M. J. (1993) "The rate of growth of phylogenetic information, and the need for a phylogenetic database." *Syst. Biol.* **42**, 562-568.

Sanderson, M. J., Donoghue, M. J., Piel, W. H. and Eriksson, T. (1994) "TreeBASE: A prototype database of phylogenetic analyses and an interactive tool for browsing the phylogeny of life." *Am. J. Bot.* **81**, 183.

Sanderson, M. J., Purvis, A. and Henze, C. (1998) "Phylogenetic supertrees: Assembling the trees of life." *TREE* **13**, 105-109.

Semple, C. and Steel, M. (2000) "A supertree method for rooted trees." *Disc. Appl. Math.* **105**, 147-158.

Strogatz, S. H. (2001) "Exploring complex networks." *Nature* **410**, 268-276.

Watts, D. J. (1999) *Small Worlds: The Dynamics of Networks between Order and Randomness.* Princeton University Press, Princeton, New Jersey.

Watts, D. J. and Strogatz, S. H. (1998) "Collective dynamics of 'small-world' networks." *Nature* **394**, 440-442.

Author's Address

William H. Piel, Institute of Evolutionary and Ecological Sciences, Kaiserstraat 63, Leiden University, 2311 GP Leiden, Netherlands.

Current Address: Department of Biological Sciences, 608 Cooke Hall, University at Buffalo, Buffalo, NY 14260, USA. Email: piel@treebase.org.

Index

www.ingramcontent.com/pod-product-compliance
Lightning Source LLC
Chambersburg PA
CBHW050551190326
41458CB00007B/1999